Henry Cantwell Wallace AS SECRETARY
OF AGRICULTURE / 1921–1924

Henry Cantwell Wallace

AS SECRETARY OF
AGRICULTURE / 1921-1924

Donald L. Winters

UNIVERSITY OF ILLINOIS PRESS

Urbana / Chicago / London

to Raena

I wish to acknowledge the help of a number of people in the preparation of the manuscript: E. David Cronon, Leland L. Sage, James W. Wallace, Donald Murphy, Donald R. Whitnah, and many at the libraries and archives where I worked. My wife Raena provided research and typing assistance, but more importantly gave encouragement at critical points along the way. A faculty research grant at the University of Northern Iowa helped to defray expenses for the typing of the final draft. *Agricultural History* and the *Annals of Iowa* have given reprint permission for material contained in chapters 2 and 6 which they had originally published.

Preface

AFTER two decades of prosperity, American agriculture in the early 1920's experienced a period of economic crisis. Emerging from World War I and the boom it generated, farmers soon found themselves confronted with a convergence of difficulties. Plummeting prices, declining purchasing power, stringent credit, increased taxes, mounting transportation costs, and a vanishing foreign market combined to visit upon the country's producers of food and fiber problems which they little understood and with which they were ill-prepared to cope. In the midst of disastrous economic conditions, farmers quickly turned to the federal government for assistance, a factor which made the post of Secretary of Agriculture in the administration of Warren G. Harding a most important and vulnerable position.

The man who took over the portfolio of Agriculture at this critical juncture was Henry Cantwell Wallace. As secretary, Wallace sought both to mitigate the immediate crisis and to place the farming industry permanently in a more favorable competitive position in relation to other sectors of the economy. The story of Wallace's work in behalf of American farmers and his efforts to bring about improvement in their industry in the early 1920's represents an important chapter in the quest for agricultural relief and reform in the United States. This book is a study of that chapter. It is not intended to be biographical and treats little of Wallace's personal life. The purpose is rather to take a close look at his tenure as Secretary of Agriculture, assess his influence over the formulation of national farm policy, and place him in the broad sweep of American agricultural history. To these ends I have attempted not only to detail Wallace's activities but also to develop the historical phenomena relevant to those activities.

Contents

Henry Cantwell Wallace AS SECRETARY
OF AGRICULTURE / 1921–1924

Chapter I / A NEW
ADMINISTRATION TAKES OVER

THE inauguration of President Warren G. Harding on March 4, 1921, marked the end of one era in the history of the United States and the beginning of another. Underlining widespread disillusionment with the previous administration, the new chief executive's lopsided victory represented a pointed rejection of Wilsonian idealism and reaction against prewar progressivism, wartime sacrifices, the fight over the League of Nations, the Red Scare, and postwar economic instability. It was a clear and unmistakable sign that Americans were a weary people, emotionally drained by the tensions of domestic and international reform, and anxious to be released from the demands of the progressive era and the empty moralizing of the Paris Peace Conference. Outside of those—more vocal than numerous—who carried into the twenties concern over the evils of alcohol and the challenges to traditional religious beliefs, the people yearned for a secure, predictable existence, free from agitation and crusades, and desired a return to a life of regularity. During the 1920 campaign, Harding caught the mood of the country perfectly, if somewhat awkwardly, with his promise of a return to "normalcy."

But whatever Americans might have wished or thought desirable in the postwar period, a return to normalcy was impossible.

For one thing, the flow of population from rural to urban areas continued to swell the country's cities, as it had in the nineteenth century. Stimulated and even encouraged during the war, business consolidation extended into a steadily increasing number of industries in the 1920's, as the giant corporation became the nerve center of the nation's economy. Technology not only continued its revolution in the factory and on the farm but moved into the home as well, with refrigerators, telephones, sewing machines, radios, and a multitude of other gadgets and appliances following in the wake of electrical power as it spread across the country. Traveling on an improved and expanding network of roads and highways, the automobile made possible a mobility undreamed of a short time earlier. If unprecedented prosperity seemed to suggest a solution to the ever-present problem of poverty, increasing maldistribution of wealth and persistent unemployment as quickly ruled out any such promise. Structural divisions within society remained sharp, although there were fluctuations as a growing group of corporation executives rose to the top of the scale and white-collar workers expanded disproportionately. And accompanying these domestic changes was the uncomfortable reality that the United States had become the greatest power and the leading creditor of the world, with all of the responsibility which its new position entailed. Normalcy, whatever that meant to Americans in 1920, was certainly not nor could it have been a characteristic of the postwar decade.

The group which sat stiffly around the large oblong table in the White House conference room at President Harding's first cabinet meeting on March 8, 1921, was hardly one prepared to meet the problems which these changing conditions were to bring. Comprising a diverse and not particularly noteworthy collection of men, some of the new advisors were holding public office for the first time, others had made inauspicious records as congressmen, while several had nothing more to recommend them for their posts than contributions to a successful presidential campaign. The shrewd and conservative businessman Andrew W. Mellon became the new Secretary of the Treasury. Crafty Harry M. Daugherty, the President's campaign manager, was rewarded with the Attorney Generalship. And Republican National Chairman Will H. Hays received the Postmaster General's portfolio. Good-natured Edwin Denby, a veteran of the Spanish-American and First World Wars and of six years in Congress, assumed the position of Secretary of the Navy, while James J. Davis, known to few outside of the Loyal Order of Moose of which he was the Director-General, took over the Labor Department. Two other former congressmen, John W. Weeks and Albert B. Fall, contributed little quality to the lackluster cabinet as Secretary of War and Secretary of the Interior, respectively.

Three members of the executive inner circle saved it from total mediocrity. Without doubt the most distinguished and in many respects the most capable of the President's recently appointed advisors was Secretary of State Charles Evans Hughes. As governor of New York, a ranking member of the Republican party, and an associate justice of the Supreme Court, Hughes had proved himself an astute political tactician and a sensitive student of human affairs. An excellent choice to direct the country's foreign relations, the new Secretary of State was Harding's most enlightened appointment. Herbert C. Hoover had likewise made an impressive record before entering the administration as director of Belgian relief and later United States Food Administrator during World War I. In both positions he demonstrated the organizational and administrative abilities which would serve him well as head of the Commerce Department in the 1920's. Rounding out the Harding cabinet was Secretary of Agriculture Henry Cantwell Wallace. Although not so widely known as either Hughes or Hoover, Wallace had gained a position of influence and respect among the country's agricultural interests as editor of the successful midwestern farm journal *Wallaces' Farmer*.[1]

Wallace was in his fifty-fourth year when he entered the Harding administration. A rounded frame of average stature, a thickset body, and a circular, boyish face topped with a full head of unruly red hair and lined with large bushy eyebrows gave him the appearance of a caricature. Pince-nez set precariously atop an angular nose revealed a pair of intense eyes and seemed to accentuate the peculiarity of his countenance. Though not a very stylish dresser, Wallace was neat and his attire was always consciously proper for the occasion. Tagged as the "dirt" farmer of the administration, the description failed to convey an accurate picture. He gave the impression instead of the small-town banker —humorless, preoccupied with his own interests, and terribly impressed with his own importance.

But beneath the outward appearance was a much different man. Confident without being vain, Wallace approached every task with an energy and a determination to see its successful completion which amazed his friends and confounded his enemies. According to his son, Henry Agard Wallace, he "went at everything with a steady intensity." [2] A practical man with an acute business sense, the new secretary possessed at the same time a set of unyielding values and a strong moral sense. When it came to fighting for principles in which he believed, Wallace reminded his close friend Gifford Pinchot of "a natural-born gamecock . . . redheaded on his head and in his soul." [3] Though certainly not deeply religious, he had a quiet Christian faith, the

---

[1] For a discussion of the Harding cabinet see Robert K. Murray, "President Harding and His Cabinet," *Ohio History* 75 (Spring–Summer, 1966): 108–25.

[2] Quoted in Russell Lord, *The Wallaces of Iowa* (Boston: Houghton Mifflin, 1947), p. 140.

[3] Quoted in *ibid.*, p. 169.

reflection of a strong Presbyterian upbringing. If not a brilliant thinker, he possessed a sound knowledge of agriculture and its problems and demonstrated a political sagacity and awareness— assets he would put to good use in his new position.[4]

Wallace's natural modesty served him well as an administrator. Instead of projecting himself into the limelight as is so often the case with men in positions of prominence and responsibility, he chose, both as farm editor and Secretary of Agriculture, to direct activities from the background and leave to others credit for accomplishments. In a profession as prestige-conscious as that of government service, it may well be imagined that this rare self-abnegation went a long way toward gaining the respect, loyalty, and gratitude of members of the Agriculture Department. Henry C. Taylor, one of his most devoted subordinates, maintained, with excusable exaggeration, that this was "one of the qualities which gave him a power of leadership and control in the Department of Agriculture unparalleled in its history."[5]

Despite a deeply ingrained sense of responsibility, Wallace had learned to enjoy life. "A curious combination of worldly impulses and a strict, high sense of duty," a man with "an Irish heart but a Scottish conscience" was the way Henry Agard Wallace described his father. Common diversions for Wallace were bowling, golf, poker, and bridge. He chewed tobacco throughout most of his waking hours and indulged, though his abstinent wife was unaware of it, in an occasional drink of bourbon or glass of beer. If he was careful to live within his means, he at the same time denied himself few of the amenities and comforts generally enjoyed by the substantial middle class to which he belonged. Freely ad-

[4] William B. Greeley, *Forests and Men* (Garden City: Doubleday, 1951), pp. 95–96.

[5] Henry C. Taylor, "A Farm Economist in Washington," p. 53 ms. in Henry C. Taylor Papers, State Historical Society of Wisconsin, Madison; and interview with Donald R. Murphy, Des Moines, Iowa, August 29, 1965.

mitting to a fondness for material things, he rejected the Spartan existence often associated with the farmers he represented in Washington. A good and frequent story-teller, he was careful that his jokes, like his attire, were proper for the occasion. Insisting that some jokes were acceptable for mixed audiences, some only for men, and still others solely for stockmen's conventions, the new secretary was always amply stocked with material to suit any group.[6]

Wallace made friends and enemies with equal ease and, it would seem, in about equal numbers. To his friends he was "Harry"—capable, sincere, faithful, and hard-working. Gifford Pinchot saw him as a man who "knows his business, is utterly honest, entirely steady and fearless, will do what he thinks is right at any cost. . . ."[7] To his enemies, on the other hand, he was unreasonable, irascible, crude, and stubborn. Herbert C. Hoover, with whom Wallace carried on a feud which began during World War I, remembered him as "a dour Scotsman with a temperament inherited from some ancestor who had been touched by exposure to infant damnation and predestination."[8] The assessments of both friends and enemies were correct. He was not one who could, or for that matter cared to, conceal his feelings. To those he liked, Wallace revealed a warm, pleasing personality; to those he disliked, a querulous, irritable disposition.

[6] Lord, *The Wallaces of Iowa,* pp. 170–72, 177, 183; Murphy Interview.

[7] Pinchot to Samuel M. Lindsay, March 6, 1921, Gifford Pinchot Papers, Manuscript Division, Library of Congress, Box 239; see also Gifford Pinchot's introduction to Henry C. Wallace's book, *Our Debt and Duty to the Farmer* (New York: Century, 1925), pp. vi–x.

[8] Herbert C. Hoover, *The Memoirs of Herbert Hoover* (New York: Macmillan, 1952), vol. 2, p. 109.

Chapter II / FROM
AGRARIAN TO AGRICULTURIST

IN one respect, Wallace's background was fitting for one selected to take over the reins of the Agriculture Department. He could count a long line of forebears who had been actively engaged in farming. In fact, his father, affectionately called "Uncle Henry" by his friends, chose a career as a Presbyterian minister and in doing so became the first Wallace of family record to break with the agrarian tradition.

The Wallaces were interested, perhaps inordinately so, in their own genealogy and kept what Uncle Henry fondly referred to as the family "Stud Book." They could trace their ancestral lineage back several generations to County Ayrshire, Scotland.[1] The first Wallaces to leave Scotland migrated to County Antrim, Ireland, in the 1680's, where Uncle Henry's father, John Wallace, was born in 1805. Joining the long stream of Scotch-Irish immigrants who settled in America, John Wallace crossed the Atlantic in 1832 and located on a farm in western Pennsylvania near the small town of West Newton. Three years later he married Mar-

[1] Material on Wallace's family background was obtained from Henry Wallace, *Uncle Henry's Own Story of His Life* (Des Moines: Wallace Publishing Company, 1917–1919); Russell Lord, *The Wallaces of Iowa* (Boston: Houghton Mifflin, 1947); and Harriet Wallace Ashby, "Uncle Henry Wallace," ms. in possession of James W. Wallace, Des Moines, Iowa.

tha Ross, the daughter of a neighboring farmer. The couple's first child was born in 1836, and in naming him they initiated a Wallace family tradition of christening the first son "Henry," which has endured for four generations.

Uncle Henry's childhood was simple, in many ways austere, but happy. Devout Presbyterians and regular churchgoers, his parents were strict and exacting in their demands upon the children. When not attending the small one-room school nearby, Henry helped his father on the 150-acre family farm or played in a wooded patch near the house. John Wallace was successively, and passionately, a Jacksonian Democrat, an abolitionist, and a Republican. As with many of similar persuasion, he was quick to form strong opinions on the questions of the day and slow to see value in views other than those he held. Henry would sit before the fire for hours and listen to his father harangue neighbors and relatives about slavery, protectionism, land policy, or whatever happened to be the current political issue or the object of his latest vexation. Witnessing these interesting and often fiery sessions, the young boy developed the militant, impulsive strain in his personality which he in turn would pass on to his first son.

Henry decided early on a career in the ministry and at the age of 18 left home to begin his training. After attending five different schools and taking time out to teach for a short while at a small Kentucky college, he completed his study in 1862. Upon receiving his license to preach, he accepted an assignment as minister of two churches, one in Rock Island, Illinois, and the other across the Mississippi River in Davenport, Iowa. Established in his new profession, Uncle Henry married Nancy Ann Cantwell, a charming little girl of Irish and Scottish descent, on September 10, 1863.

Uncle Henry was jovial, good-natured, and made friends easily. But at the same time he was opinionated, dominating, and had a penchant for alienating those with whom he disagreed. A highly articulate man, he was at his best delivering an extemporaneous sermon. Although a devout Christian with strong Puritan roots, he was able to rise above the confining dogma of nineteenth century Presbyterianism. Possessed of a hedonistic nature, he derived a true joy from life and had a fondness for material, as well as spiritual, comforts. Nancy Cantwell Wallace was decidedly Irish in temperament. Petite, perhaps on the frail side, she would have given the impression of being physically weak had it not been for a head of fiery red hair which suggested an inner vitality. She was an intense, spirited woman, an immaculate housekeeper, and a devoted mother and wife. Life with Uncle Henry was at times trying, however, for his carefree nature was understandably irritating to one so exacting as she. Like her husband, Nancy Cantwell Wallace was deeply religious without being pious.

The couple began married life in a small Rock Island apartment, Uncle Henry busy with his religious duties and Nancy with the responsibilities of a new wife. As an idealistic and rather aggressive young minister, Henry soon instituted several changes in the church service and policy. Although the traditionalists of the Rock Island and Davenport congregations opposed these innovations, often vehemently, their stubborn spiritual leader held his ground and refused to yield. The result was a state of religious turmoil and mutual resentment. It was into such an atmosphere that Henry Cantwell was born on May 16, 1866.

When young Henry was five years old his father secured a position with the United Presbyterian Church of Morning Sun, Iowa. Tired of the constant bickering between his two congregations in Rock Island and Davenport over where he should live, how his time was to be divided, and a multitude of other petty differences, Uncle Henry welcomed the chance to serve a single flock and quickly accepted the call proffered him in 1871. The move to Morning Sun was a fortunate one for the Wallace family. Uncle Henry's health, which had been declining steadily in Rock Island, soon showed signs of recovery. The small-town, rural atmosphere of Morning Sun appealed to both him and his wife far more than the faster pace of a river port like Rock Island. Naturally gregarious and friendly, the recent arrivals were not long in assuming a place of respect and prominence in Morning Sun. It was here, too, that young Henry attended his first school, and if not the top student in class he at least did reasonably well in his studies.

Soon after settling in Morning Sun, Uncle Henry took an active interest in politics. Like his father, he had become a staunch Republican with the approach of the Civil War and tenaciously supported the party in the post-bellum era. As with many in the latter nineteenth century, he was particularly concerned over the money issue. Resolutely opposed to greenbacks and later to free silver, he favored a "sound" money system based on a gold standard. Although a supporter of the high tariff policy of the Republican party in the immediate post-Civil War period, Uncle Henry was to change his views on that question after he became personally involved in the fight to improve agricultural conditions. His one excursion as a political candidate came in 1875 when he ran unsuccessfully for a seat in the Iowa Senate. The defeat failed to dampen his interest and enthusiasm for politics, however, and he remained an active supporter of the Republican party until his death in 1916.

Uncle Henry's health declined again in 1877, and on the advice of his physician he left the strenuous life of the ministry. After moving his family to the small southwestern Iowa town of Winterset, he engaged in light farming on some land he had purchased earlier in nearby Adair County. A short time later he established a creamery in Winterset, supplied in part by his own herd of Jersey cows. But the venture lost money from the start and was soon closed down. Uncle Henry, it was obvious, had little business acumen. Then, in 1879, the editor of the *Madisonian,* Winterset's leading newspaper, approached Uncle Henry with an offer to write a weekly agricultural column. Although the retired minister could boast no journalistic experience, he had maintained a lively interest in agriculture since his childhood days on his father's farm in Pennsylvania and the proposition appealed to him. In accepting the job, he began the development of an interest and ambition that would eventually lead to the creation of one of the country's most influential farm journals.

Disagreement with the editor over certain political opinions expressed in the column, however, rendered Uncle Henry's association with the *Madisonian* a short one. Encouraged by the favorable public acceptance of his articles, he decided in 1880 to purchase half interest in a failing Republican sheet, the *Winterset Chronicle,* and begin publishing his own paper. The newspaper was an immediate success, so much so that within a year Uncle Henry had bought complete control of the *Chronicle* and its town rival, the *Madisonian.* Soon the new editor-publisher was gaining a state-wide reputation not only as a journalist but as an expert on farm problems.

The interest aroused by Uncle Henry's agricultural writings resulted in his appointment in 1883 as editor of the farm journal *Iowa Homestead.* Although the office of the *Homestead* was located in Des Moines, Uncle Henry was permitted to remain in Winterset and send his material in to the paper. Two years later James M. Pierce purchased control of the journal, and he retained Wallace as his editor. The change in ownership brought about the association of two men who would eventually become bitter rivals as the state's leading agricultural journalists. For 10 years, however, they were to work together, with Wallace handling editorial policy and Pierce acting as business manager, to build the *Homestead* into one of the outstanding farm journals of the Middle West.

At the time his father was rising to eminence as a farm editor, Henry Cantwell Wallace was growing up in the small-town environment of Winterset. The county seat of Madison County with a population of 2,600, Winterset was set in the middle of a thriving agricultural region devoted to livestock and grain production. As the terminus for one branch of the Rock Island Railroad, the town served as a marketing and business center for the surrounding area. Compared to other nearby settlements, it had good schools, adequate retail and service establishments, and was generally prosperous. Typical of many of the rural-oriented communities scattered throughout the developing West, Winterset was relatively free of class distinctions and inhabited by friendly, sociable people with little diversity in outlook or attitudes.

Young Henry's life was routine, generally unexciting, and taken up largely with school and work. At home he lived within an atmosphere of agricultural journalism, listening to Uncle Henry expound on the farm situation, read aloud drafts of forthcoming articles and editorials, and complain of a multitude of problems connected with the publishing business. Working alongside of his father, he spent much of his time outside of school setting type, printing, soliciting advertising, and delivering finished copies. The journalistic experience young Wallace gained and the interest he developed at an early age would be strong influences in shaping his later life.

With his father busy on the *Homestead,* Henry graduated from high school and decided to go on to college. In 1885 he entered Iowa State Agricultural College (now Iowa State University) at Ames and began preparation for a career in agricultural teaching and research. Diligent and hard-working, he did well in his studies, which included heavy concentrations in both science and mathematics.[2] His primary interest, however, was in a student organization called the Agricultural and Horticultural Associa-

[2] Wallace's College Record, Iowa State College, Ames.

tion and in a small paper it published, the *Student's Farm Journal,* devoted largely to reporting the activities of the agriculture department and extension service of the college. As a member of the editorial board, Wallace further developed his journalistic interests.[3]

Without a clear idea of what he wanted from school or of the kind of career he wished to pursue, Henry found himself bored with college. Thus when a tenant left one of Uncle Henry's farms, the young man asked if he might take over and operate the property. Reluctantly his father agreed, and at the end of his sophomore year Henry quit school and moved onto the farm in Adair County, Iowa.[4] Life in the country was lonely, particularly for a bachelor, and it was not long before young Wallace began to contemplate marriage. He had courted an attractive coed at Ames and had found her to be quite compatible. Never one to delay long in making important decisions, Henry proposed in the summer of 1887 and May Brodhead, herself not one to delay, promptly accepted. The young couple—she was 20 and he was 21—were married on Thanksgiving Day, November 24, 1887.

Of English, Dutch, and French Huguenot descent, the new bride was born in New York City.[5] Her parents had died before she was old enough to remember them, however, and she had gone to live with an aunt in Muscatine, Iowa. From a family of strong Republicans and even stronger Methodists, May Brodhead was taking an unlikely degree in art and music at Ames

[3] Earle D. Ross, *A History of Iowa State College of Agriculture and Mechanic Arts* (Ames: Iowa State College Press, 1942), p. 179.

[4] Wallace to Clarence Poe, May 27, 1924, Files of the Secretary of Agriculture, National Archives. Uncle Henry began speculating in land in 1869 and had acquired three farms by 1887, all of which he leased to tenants. *Tributes to Henry Wallace* (Des Moines: Wallace Publishing Company, 1919), pp. iii–iv.

[5] "Ancestral Tabulation," ms. in Henry C. Wallace Papers, University of Iowa Library, Iowa City.

before marriage interrupted her college training. A slender, but sturdy, woman of average height, she was neat in appearance and an interesting contrast to her stocky husband. Sensitive and headstrong, she was one of the few in the Wallace clan who could successfully stand up to the patriarchal Uncle Henry. When the old man became pompous beyond endurance, she would turn to him and firmly assert, "I'm not a Wallace; I only married one." [6]

The life of a farm wife must have taken considerable adjustment for May Brodhead Wallace. She had never before lived on a farm and knew little of agriculture, which would have been problem enough; but to complicate matters further, the 1880's were not a particularly favorable time for a young couple to enter farming. Unstable economic conditions, sharply fluctuating prices, and uncertain profits made agriculture a precarious occupation in the post-Civil War period. Farming demanded long hours and hard work and even then could not assure a reasonable return on labor and investment. But Mrs. Wallace bridged the gap between coed and farm wife with a minimum of difficulty. That adjustment to the problems of rural life had come rather rapidly she demonstrated on the birth of her second child, Annabelle. Returning from town with the doctor, Wallace found that their arrival was tardy, for the baby had already been born and was perched on her mother's knee.[7]

In early 1892 James Wilson, soon to become Secretary of Agriculture in the William McKinley administration, took over as head of the department of agriculture at Ames. Mutual interest in farm problems had led to a warm friendship between him and Uncle Henry. Wilson had followed the career of his friend's son and regretted Henry's decision to drop out of school and take up

[6] Quoted in Lord, *The Wallaces of Iowa,* pp. 106–7.
[7] Interview with James W. Wallace, Des Moines, Iowa, December 21, 1966.

farming. After his appointment as department head, he offered young Wallace a position as assistant professor of dairying at Ames if he would return to college and complete his degree in agriculture. Henry decided to accept the offer, and the spring of 1892 found him, along with his wife and two children, Henry Agard and Annabelle, back in Ames.[8] On his return to college Wallace was much better adjusted, more mature, and surer of his plans than he had been in 1885. Completing the final two years of course work in a single academic year, he graduated with a bachelor's degree in December 1892 and received appointment to the staff of the school's agriculture department at a salary of $1700 per year.[9]

At the time Wallace joined the Iowa State Agricultural College faculty, the agriculture department had few courses in dairying, and it was his job to broaden the offerings. He was particularly interested in the field of dairy bacteriology and sought to develop the departmental curriculum along these lines. But, although he taught one class in dairy bacteriology, objections from members of the science faculty, who felt that such courses should be offered by their department, hampered his efforts.[10] Still, Wallace succeeded in strengthening his division, and, according to Wilson's assessment at the time, "the facilities for teaching dairying in a thoroughly practical and scientific manner are unexcelled by any college in the country." [11]

Shortly after accepting his position at Iowa State, Wallace and

[8] Lord, *The Wallaces of Iowa,* p. 118; and Harlan Miller, "Secretary of Agriculture Represents New Aristocracy of the Soil," *Des Moines Register,* February 27, 1921, Magazine Section, p. 2.

[9] Report of the Board of Trustees of Iowa State College, 1892, p. 441; and Wallace to Clarence Poe, May 17, 1924, Files of the Secretary of Agriculture.

[10] F. A. Leighton to E. W. Stanton, February 27, 1920, and L. H. Pammel to George L. McKay, March 1, 1922, Iowa State History Collection, Dairy Industry Department, Iowa State University Library, Ames.

[11] Louis H. Pammel, *Prominent Men I Have Met: Henry Cantwell Wallace* (Ames: Powers Press, n.d.), p. 6.

another member of the Ames staff, Charles F. Curtiss, purchased a little-known livestock journal published in Cedar Rapids called the *Iowa Farmer and Breeder*. Begun in 1875, the sheet had been acquired in 1891 by Wallace's brother-in-law, Newton B. Ashby. Ashby won appointment to a consulate in Ireland two years later, however, and was forced to sell the journal. Little did Wallace realize it at the time, but purchase of the journal by the two Ames agriculturalists was an important step toward the eventual establishment of *Wallaces' Farmer*.[12] They moved the paper to Ames and renamed it *Farm and Dairy*. Henry's younger brother, John P. Wallace, was brought into the small organization, sold a share of the paper, and placed in charge of advertising. Published semimonthly, the sheet was devoted primarily to reporting and promoting the dairy work of the agriculture department and experiment station of the college. Its circulation small and profit negligible, the journal nevertheless began to draw the attention of farm interests in the state.[13]

Limited as its circulation and objectives were, *Farm and Dairy* became involved in a controversy scarcely a year after the new ownership took over. The incident arose when Wallace and Curtiss questioned the integrity of an experiment station chemist, G. E. Patrick, and the validity of some of his findings. In a letter to the Board of Trustees of the college, *Farm and Dairy* demanded an investigation and threatened to publicize the issue if appropriate action were not taken. After considering the charges, the board held them to be groundless and refused to pursue the matter. It also adopted a resolution declaring that resignation would

[12] Gerald L. Seaman, "A History of Some Early Iowa Farm Journals," Master's thesis, Iowa State University, Ames, 1942, pp. 52–54; and Earle D. Ross, *Iowa Agriculture: A Historical Survey* (Iowa City: State Historical Society, 1951), p. 89.

[13] Arthur T. Thompson, "Wallaces' Farmer," *Palimpsist* 11 (June, 1930): 245–46; and Wallace, *Uncle Henry's Own Story*, vol. 3, pp. 70–71.

be demanded of any Ames professor publishing or supplying for publication criticism of another staff member.[14]

With the board's failure to act, *Farm and Dairy* carried out its threat by publishing its charges against Patrick in May 1894. Immediately the article aroused a controversy within the agriculture department and drew protests from certain livestock and dairy interests who supported the work of the experiment station. The station and its backers retaliated by accusing Wallace of taking credit for development of an improved method for extracting butter fat when in fact it had been discovered by someone else. Since the college was "being materially injured by these dissensions," the Board of Trustees decided at its next meeting in August 1894 that, "in the interest of harmony," it would request resignations from Wallace, Patrick, and two other professors who had become involved in the conflict and order Curtiss to sever his connections with *Farm and Dairy*. On the same day Wallace tendered his resignation, effective at the end of the current term in March 1895.[15]

At the same time the Ames controversy was taking place, Uncle Henry was having his trouble with James M. Pierce over the editorial policy of the *Iowa Homestead*. Under Wallace's direction the journal had taken a militant antimonopoly stand; in fact, according to Uncle Henry, the *"Homestead* became more of an anti-monopoly paper than an agricultural one."[16] Early in 1894 the railroads announced their intention to raise freight rates, and Wallace, consistent with established policy, proposed a propaganda campaign against the planned increase. Opposed for some reason to such action, Pierce prevented publication of an article which his editor had written to initiate the campaign, and a bitter argument ensued. With the controversy still raging,

[14] Report of the Board of Trustees of Iowa State College, 1894, pp. 549–54, 569–75.

[15] *Ibid.,* 1895, pp. 11, 23.

[16] Wallace, *Uncle Henry's Own Story,* vol. 3, p. 52.

Uncle Henry left on a trip to Europe, only to learn upon his return that Pierce had removed him as editor of the *Homestead*.[17]

By early 1895, then, both father and son found themselves out of jobs, which, as it turned out, was most fortunate. The solution to their mutual problem, they discovered, was really very simple. Why not purchase Curtiss's share of *Farm and Dairy,* of which he had been ordered to divest himself, and establish an agricultural journal of their own? Curtiss was glad for the chance to rid himself of his troublesome piece of property, particularly since he had ambitions (never realized) of becoming president of Iowa State. With the final transfer in February 1895, the Wallace family paper had its beginning.[18]

Somewhat sentimentally, young Henry would remember years later that the family had embarked upon this undertaking because of the belief that "father would die if he could not have a platform to stand on and continue his work." [19] He recalled that "it did not seem a very hopeful prospect, but on father's account it was the only thing to do." [20] Fortunately for *Wallaces' Farmer,* young Henry's business sense was better than his memory. First of all, the Wallace family was too shrewd to set out on so uncertain a venture merely to satisfy the whims of even so lovable a man as Uncle Henry. By 1895, furthermore, he was widely known and respected as a farm editor and would have had little problem securing another position after his severance from the *Homestead*. It seems unlikely, therefore, that the Wallaces had nothing more in mind than a vehicle of expression for the head of the clan.

The Wallaces were in a poor financial position for the under-

[17] *Ibid.,* p. 68; Seaman, "A History of Some Early Iowa Farm Journals," p. 39; and Ashby, "Uncle Henry Wallace," p. 56.

[18] Lord, *The Wallaces of Iowa,* p. 130; and Earle D. Ross to the author, October 8, 1966.

[19] Quoted in Henry C. Taylor, "Henry C. Wallace," ms. in Henry C. Taylor Papers, State Historical Society of Wisconsin, Madison, p. 1.

[20] "Statement of H. C. Wallace," ms. in possession of James W. Wallace, Des Moines, Iowa, p. 1.

taking of such a project. With most of Uncle Henry's money tied up in *Homestead* stock, of which he could not dispose, they had no cash reserves. Thus they were forced to obtain loans wherever possible: Curtiss accepted a mortgage on his share of the journal; James Wilson signed a $5000 note at a bank which he renewed each year until 1900; Uncle Henry borrowed from a brother; young Henry mortgaged his home in Ames for $1500 and got another $5000 by using his life insurance policies as collateral.[21] To make matters worse, the 1890's was a precarious time to begin an agricultural journal. Facing an even more serious economic situation than Henry had experienced during his short try at farming in the previous decade, producers frequently found it difficult to spare the fifty cents for a year's subscription to *Farm and Dairy*. Unable to sell their products to destitute farmers, advertisers were reducing their space in established journals and were thus reluctant to buy from a new paper.

The journalistic experience and talents of Uncle Henry coupled with the business ability of his oldest son, however, permitted the sheet to survive the first lean years until circulation and advertising had been built up. Also contributing to the sheet's early success was the fact that *Farm and Dairy* from the beginning was a family project, and the Wallaces hired no more outside help than was absolutely necessary. Uncle Henry was the editor and young Henry the general manager; John continued to be in charge of advertising; the youngest son, Daniel, set type and ran errands; and Nancy Wallace wrote a column on homemaking. By the end of seven years all debts had been paid and the journal was clear.[22]

A successful tactic used by the new paper to increase circulation was raiding subscribers of the *Homestead*. In the issue of February 15, the first one after the Wallaces took over *Farm and*

[21] *Ibid.*
[22] *Ibid.*, p. 2; and Taylor, "Henry C. Wallace," p. 1.

Dairy, and in every number for several years thereafter, they placed an announcement at the top of the first page explaining why Uncle Henry had left the *Homestead* and soliciting the support of old friends who had followed his editorials in that journal.[23] This bit of strategy, plus the fact that the two papers engaged in a running public feud for many years, helped to bring a steadily increasing circulation in the latter nineteenth and early twentieth centuries. Continuing after the deaths of Uncle Henry and James Pierce, the fight between *Wallaces' Farmer* and the *Homestead* lasted until 1929 when, ironically, both journals faced financial difficulties and were forced to merge into *Wallaces' Farmer and Homestead.*

The latter 1890's saw a number of changes in the young journal. On March 1, 1895, the first issue appeared under the new name of *Wallaces' Farm and Dairy.* Six months later the name was changed again to *Wallaces' Farmer and Dairyman,* with weekly rather than semimonthly publication. Early in 1896 the paper moved from Ames into its new office in Des Moines, Iowa. And finally, in December 1898, the journal appeared as simply *Wallaces' Farmer,* the name it would retain until the merger with the *Homestead* in 1929. During the first few years, circulation mounted, advertising space increased in value, and the paper grew in size. By the turn of the century *Wallaces' Farmer* was well established and had marked out for itself a place among the influential agricultural journals of the Middle West.[24]

[23] *Farm and Dairy* 20 (February 15, 1895). The announcement read: "Mr. Wallace was for 10 years the editor of the *Iowa Homestead.* His withdrawal from that paper was the culmination of trouble between him and the business manager as to its public editorial policy, Mr. Wallace wishing to maintain it in its old position as the leading western exponent of anti-monopoly principles. Failing in this he became the editor of the *Farm and Dairy,* over the editorial policy of which he has full control. He invites the co-operation of his old *Homestead* friends in making *Farm and Dairy* the leading western authority on agricultural matters."

[24] Thompson, "Wallaces' Farmer," p. 253; and Seaman, "A History of Some Early Iowa Farm Journals," p. 39–40.

Meanwhile, Uncle Henry remained active in the Republican party and had risen to a place of importance in the state organization. That he had attained a position of political influence not only in Iowa but at the national level as well was testified to in 1896 when President-elect William McKinley consulted him in regard to possible candidates for the post of Secretary of Agriculture. No doubt Uncle Henry's strong recommendation of his close friend James Wilson was in part responsible for the Ames professor's subsequent appointment.[25] Further evidence of Wallace's position both in the Republican party and as a recognized authority on agriculture came during the administration of Theodore Roosevelt. In 1908 the President appointed Uncle Henry to the newly created Country Life Commission, the purpose of which was to investigate rural culture and life and to make recommendations for their preservation and improvement.[26]

Despite the fact that both Uncle Henry and his oldest son were strong and active Republicans, they attempted to avoid party politics in the columns of their farm journal. *Wallaces' Farmer* supported or opposed specific measures and took positions on single

[25] Willard L. Hoing, "James Wilson as Secretary of Agriculture, 1897–1913," Doctoral dissertation, University of Wisconsin, Madison, 1964, pp. 14–15; and Wallace, *Uncle Henry's Own Story*, vol. 3, p. 80. An often repeated, but totally undocumented, story holds that McKinley initially offered the portfolio of the Secretary of Agriculture to Wallace, who declined and recommended Wilson instead. While it can be neither proved nor disproved, it seems quite likely that the account is a product of state chauvinism. Iowans delight in telling that Wallaces of three successive generations were offered the position as head of the Agriculture Department, the last being Henry Agard who was secretary in the Franklin D. Roosevelt Administration. James W. Wallace, Henry Cantwell's son, and Donald R. Murphy, a longtime friend and associate of the Wallace family, both maintain that it is doubtful if Uncle Henry were ever offered the position. Russell Lord, biographer of the Wallaces, also casts doubt on the validity of the story. Wallace Interview; Interview with Donald R. Murphy, Des Moines, Iowa, August 29, 1965; and Lord, *The Wallaces of Iowa*, pp. 133–35.

[26] Wallace, *Uncle Henry's Own Story*, vol. 3, pp. 100–102.

issues, but the paper's policy was never to endorse any candidate for public office nor to support a particular political party.[27]

Following a nonendorsement policy was often difficult for two men of such strong political views and militant temperaments, however, and on a few occasions they violated the spirit, if not the letter, of their professed practice. The presidential election of 1912 was a case in point. Personal friends and ardent admirers of Theodore Roosevelt, the two Wallaces had firmly supported his policies as President. As early as 1910 Uncle Henry was criticizing Roosevelt's successor, William Howard Taft, for unwittingly becoming an instrument of special interests. He praised Republican insurgents for their opposition to business domination of the party and endorsed their "stand for Rooseveltism as he himself understood it." [28] When Roosevelt bolted the Republican party and formed his own Progressive party in 1912, young Henry and his father, in one of the few times either of them failed to support an official Republican candidate, followed their progressive idol. During the campaign, *Wallaces' Farmer* adhered narrowly to a nonendorsement policy, but its editorials were highly favorable to Roosevelt's candidacy and laudatory of the Progressive party. While approaching the election "from a philosophic and stricty non-partisan standpoint," the journal recommended that readers vote for the "man" and not hesitate to cross party lines. The Wallaces advised the electorate to judge each candidate on his past record and then to support "whichever one seems to you to present the best policy for the whole people." Given his popularity, reputation, and third-party affiliation in 1912, such counsel was obviously calculated to redound to Roosevelt's advantage. *Wallaces' Farmer* further defined its concept of "strict non-partisanship" by observing that the Pro-

[27] *Wallaces' Farmer* 35 (June 6, 1910): 643.
[28] Henry Wallace, "What the Middle West Wants," *World's Work* 20 (May, 1910): 12897.

gressive party "offers a political home to both republicans and democrats who believe in what we have come to call progressive principles." [29]

In addition to his work on the paper and in the Republican party, young Henry was also active in the Corn Belt Meat Producers' Association. Formed in 1905 largely through the efforts of Wallace, the organization had as its principal objective the reduction of railroad rates which discriminated against Iowa livestock shippers. As would be his administrative method as Secretary of Agriculture, Wallace chose to work in the background but at the same time to retain firm control over the association. From his position as permanent secretary of the group, he picked the president, vice president, and members of the board of directors and for the most part dictated policy. A. Sykes, long-time president of the organization, did much as he was told by Wallace and later by Henry Agard, who took over the secretary's duties after his father became head of the Agriculture Department. [30]

Although small, the Corn Belt Meat Producers' Association represented the biggest livestock feeders in the state and was one of the most active organizations of its kind in the early twentieth century. It hired an aggressive young attorney, Clifford Thorne, and trained him as a rate expert. Thorne and Wallace made several trips to Washington to represent organization members before the Interstate Commerce Commission, and in at least one case they were instrumental in the commission's refusal to grant an increase in freight charges on stocker and feeder cattle. The Corn Belt Meat Producers' Association also appeared regularly at hearings held by the Iowa State Railroad Commission. [31] Al-

[29] *Wallaces' Farmer* 37 (October 11, 1912): 1426 and 37 (August 16, 1912): 1166.

[30] Murphy Interview.

[31] *Ibid.;* Wallace to W. G. Alcorn, March 26, 1912; Wallace to George Kruise, Jr., November 9, 1911; Wallace to Board of Railroad Commis-

though it was the group's policy to avoid involvement in contests for state offices, it worked behind the scenes to influence the appointment of the Railroad Commissioner.[32] Not primarily a political pressure group, the association still petitioned congressmen and solicited their support for or opposition to legislation affecting transportation conditions.[33] As secretary, Wallace also corresponded with railroad owners demanding such things as better sleeping accommodations for shippers, more careful handling of livestock in shipment, and improved facilities for loading and unloading. In several instances the association represented individual members in making claims against railroads for damages resulting from negligence or overcharging.[34]

In time the interests of the association broadened to include not only matters of transportation but those involving agricultural interests in general. Thus by 1919 it was working for federal regulation of the meat packing industry and commodity exchanges, demanding greater emphasis on farm economics in the country's agricultural colleges, and supporting the federation of county farm bureaus into state and national units.[35] The expanded program of the Corn Belt Meat Producers' Association reflected Wallace's growing concern for and awareness of the problems which were increasingly plaguing the nation's farmers.

sioners of Iowa, April 29, 1912; and James M. Brockway to Wallace, April 2, 1912, Wallace Papers, Box 1.

[32] Wallace to J. C. Edwards, May 12, 1912; and Wallace to James M. Brockway, May 24, 1912, *ibid.*

[33] Wallace to H. E. Kendall, February 8, 1911; Wallace to Albert B. Cummins, March 6, 1912; Wallace to Albert B. Cummins, August 12, 1912; and Wallace to N. S. Ketchum, January 25, 1912, *ibid.*

[34] There is much correspondence pertaining to shipping conditions in the Wallace Papers. See for example Wallace to R. H. Aishton, January 25, 1912; Wallace to B. L. Heatherton, April 5, 1913; Wallace to C. A. Baker, January 12, 1912; and Wallace to Directors of the Association, October 18, 1920. For claims action see Wallace to Pat Kelly, November 21, 1911; Wallace to H. O. Allison, March 11, 1912; and Wallace to H. S. Boomgaarden, March 19, 1912, *ibid.*, Boxes 1 and 2.

[35] Wallace to John B. Kendrick, February 2, 1920, *ibid.*, Box 1; and *Wallaces' Farmer* 44 (February 7, 1919): 322.

After a full, rewarding, and successful life, Uncle Henry died on February 22, 1916, at the age of 79. Although saddened by the death of his devoted Nancy seven years earlier, he remained active up to his last days as editor of *Wallaces' Farmer*. That he had built an admirable reputation as an agricultural expert was attested to by the hundreds of tributes he received after his death. Arthur Capper, himself a farm editor of repute, pronounced Uncle Henry "the dean of agricultural journalism." Bradford Knapp, son of Seaman Knapp who popularized farm demonstration work, regarded him as "a true agricultural statesman," while a county extension agent in Illinois considered him "the greatest agricultural editor of his day" and believed that "no other man has won the hearts of the farm people in the same way that Henry Wallace did." And his good friend Gifford Pinchot thought he was "one of the two or three best men I ever knew."[36] Uncle Henry had made enemies, and probably deserved many of them, but he had also served agriculture in an important way. By calling attention through his editorials to the plight of a segment of society which a rapidly industrializing country had come to ignore, he contributed to a growing awareness of the problems of farmers which would someday result in responsible action.

With the death of his father, Henry Cantwell Wallace took over as editor of *Wallaces' Farmer*. He had played an integral part in the paper's rise to a place among the top farm journals of the Middle West and in doing so had learned the business well. This experience, coupled with a journalistic ability which he had acquired through hard work rather than natural endowment, enabled him to sustain the high quality readers had come to expect from *Wallaces' Farmer*. And despite the new editor's fear that circulation might decline after Uncle Henry's death, the sheet not only held its own but strengthened its position.[37]

[36] *Tributes to Henry Wallace*, pp. 158, 185, 201, 213.
[37] "Statement of H. C. Wallace," p. 2.

The format of the paper continued much the same as it had been under Uncle Henry's direction. Each issue contained a number of editorials on subjects of timely interest to the rural community. Frequently of a controversial nature, they attacked business and financial interests, agitated for passage or repeal of measures before Congress, urged stronger government action to assist farmers, and generally maintained the same editorial tone established by Uncle Henry. Feature articles dealt with practical problems and kept readers abreast of the latest agricultural developments. Serialized stories, which substituted action for literary merit, were an added attraction. One of the most popular sections was "Uncle Henry's Sabbath School Lesson," a weekly account of a biblical story in simple, clear prose. Uncle Henry had initiated the series a few years after the establishment of *Wallaces' Farmer,* and it proved so successful that he prepared an advance supply of weekly articles to last for 21 years. After his death, the journal continued to publish the articles on a rotation system which Uncle Henry had set up. Also retained as a regular part of the paper was the column on homemaking and household hints begun by Nancy Wallace. In addition, the journal included reports on farm organization conferences, state fairs, activities of the agricultural college at Ames, and the work of the Iowa state extension service.

Continuing the tradition of a family paper, Wallace brought his own children into the organization. Upon graduation from Iowa State College in 1910, Henry Agard, his oldest son, became a member of the editorial staff. While at Ames he had developed an interest in corn breeding, and besides writing for *Wallaces' Farmer* he continued his experiments in hybrid varieties. Eventually this work was to contribute to the development and ultimate widespread use of hybrid seed corn and to lead to the establishment of the Pioneer Hy-Brid Corn Company in 1926. Just as his father had before him, Henry Agard learned the journalistic trade on *Wallaces' Farmer,* and when the elder Wallace left to become Secretary of Agriculture in 1921 he assumed active editorship of the paper. Two other sons, both graduates of the University of Pennsylvania, were later to join the journal—John entering the advertising department and James working in an administrative position. After her graduation from Vassar, Annabelle, one of Wallace's three daughters, became an editorial writer.[38]

In addition to his duties as editor and as secretary of the Corn Belt Meat Producers' Association, Wallace found time for other activities. He was a thirty-second degree Mason, belonged to the Des Moines Club, and was a member of the Prairie Club, an informal organization which met once a month to hear and discuss a paper written by one of the group. For many years he was active in the Young Men's Christian Association, serving as chairman of the Des Moines unit and during World War I as a member of the organization's National War Council. Though he lacked Uncle Henry's strong religious orientation, Wallace regularly attended the United Presbyterian Church in his hometown and later the New York Avenue Presbyterian Church of Washington, D.C. Always interested in agricultural education, he maintained an unofficial contact with Iowa State College. Many of his suggestions on curriculum and policy, such as the need for a broader education for students in agriculture, however, met with poor reception in his old department. "At the time I was in charge of the Dairy Department," recalled Wallace's successor at Ames, "Henry Wallace seemed to have the faculty of getting into trouble with a number of professors there." [39]

There was also time for recreation and enjoyment. Although he began the game late in life, Wallace became an avid golfer

[38] Wallace Interview.

[39] Wallace to A. B. Storms, November 5, 1909; Storms to Wallace, November 3, 1909; and G. L. McKay to L. H. Pammel, March 3, 1922, Iowa State History Collection, Dairy Industry Department.

and was frequently seen on the Country Club course. With his wife and family, he enjoyed bowling and horseback riding. Perhaps more than anything, the Wallaces liked having friends into their home for bridge, impromptu dancing, or just conversation. In the years before leaving for Washington, they entertained such well-known guests as Theodore Roosevelt, William Howard Taft, John D. Rockefeller, Jr., Gifford Pinchot, and Ray Stannard Baker.[40]

Understanding, but at the same time firm with his six children, Wallace was a fine parent. Within a busy schedule he always made certain that time was set aside for his family. An incident involving James, the youngest son, well illustrates the kind of relationship that he developed with his children. As boys of 14 are wont to do when they are allowed to drive the family car alone, James was tempted into a race with a companion. Rounding a corner the high-top Franklin automobile tipped onto its side, causing considerable damage but no injury. Upon being told of the accident, Wallace merely suggested that he and James should retrieve the car and said nothing more about the misfortune.[41]

As for most Americans, World War I was a rather trying period for Wallace. Farmers, to be sure, were enjoying a heady prosperity as prices for their products soared. But there were accompanying problems and irritants, of which one of the most serious was the government's handling of wartime production and distribution of agricultural goods. With the entry of the United States into the war, President Woodrow Wilson placed Herbert C. Hoover at the head of the newly created Food Administration, which was charged with responsibility for formulating and carrying out the federal program. Hoover in turn appointed Wallace as chairman of a committee to advise the new agency on pork supply.

[40] Wallace Guest Book in possession of James W. Wallace, Des Moines, Iowa; and Wallace Interview.
[41] Wallace Interview.

Disagreement soon developed over policies designed to stimulate hog production and led to a bitter conflict between the future cabinet colleagues.[42]

When it became clear in the latter part of 1917 that the supply of pork was failing to meet mounting wartime demands, the Food Administration attempted to correct the deficiency through a propaganda campaign. Basing its approach on an appeal to rural patriotism, Hoover's office issued circulars and press releases to the effect that it was the farmers' duty to increase output. "Closing the gap in hog production . . . ," read one release, "is not only one of the big opportunities but one of the big obligations of American farmers. . . . This is the immediate war duty of farmers." [43] Along with efforts to increase production, the Food Administration also admonished the country as a whole to consume less meat.

Wallace, on the other hand, insisted that the approach of the Food Administration would not attain the desired result. "Appeals to the producer on the ground of patriotism," he lectured Hoover, "will not bring an adequate response." Convinced that production could be stimulated only by an "economic appeal," Wallace urged a solution based on a theory worked out by his oldest son.[44] From a study he had made of the relationship between livestock and grain prices, Henry Agard concluded that the ratio of the market offering per hundredweight of hogs and that per bushel of corn had to equal or exceed 13:1 if production of pork were to be stimulated to the degree needed in 1917. As long as hog prices remained low in relation to corn, he explained, most farmers would find it more profitable to sell their

[42] For a complete discussion of the controversy, see Donald Winters, "The Hoover-Wallace Controversy during World War I," *Annals of Iowa* 39 (Spring, 1969): 586–97.

[43] Undated Press Release of the Food Administration, copy in Wallace Papers, Box 4.

[44] Wallace to Herbert C. Hoover, September 24, 1917, *ibid.*, Box 1.

corn rather than to feed it to livestock, and pork production would thus remain below the desired level.[45] The "economic appeal" advanced by the older Wallace was one in which the Food Administration would guarantee hog prices based upon the formula derived by his son.

Hoover had little interest in the proposal and set himself against it from the outset. As pork production remained below the needed level, however, pressure mounted for a change in the Food Administration's policy, and he finally appointed a commission, recommended by Wallace's committee, to study the price-ratio plan. After a favorable report by this group, Hoover's office announced in November 1917 that it would use its influence over the market to "try" to stablize the price of hogs farrowed in the spring of 1918 at the 13:1 ratio.[46] But when these animals began to enter the market in the following September, the Food Administration first applied the ratio formula in a way which was unsatisfactory to producers and then decided to abandon it entirely.[47]

Wallace was furious. He charged that farmers had increased hog production in response to the Food Administration's promise only to find that there was no intention of honoring the commitment. On the other hand, Hoover complained that farmers had

[45] Henry A. Wallace, *Agricultural Prices* (Des Moines: Wallace Publishing Company, 1920), pp. 28, 33–35. After the war, Wallace published his theory of the ratio method of determining prices in this book. Subsequent reference to the hog-corn ratio will always mean the relation between the price of hogs per hundredweight and that of a bushel of corn. The grade of hogs will be understood to be good to select.

[46] Advisory Commission to Hoover, October 27, 1917, copy in Wallace Papers, Box 1. The report was published by the Food Administration in a pamphlet entitled *Hog and Corn Ratios* (1917); and *Prices of Hogs* (United States Food Administration, 1917).

[47] Press releases of the Food Administration, September 25, 1918 and October 26, 1918, copies in Wallace Papers, Box 4. Producers were unhappy with the Food Administration's decision to use the price of corn in the country markets rather than the higher Chicago price as the basis of its policy.

misinterpreted the policy of his office as an absolute price guarantee rather than the statement of intention which it was. As for the decision to drop the policy, he explained that this had been done because of the sharp decline in corn prices with the approach of the end of the war, insisting that the ratio method would thus have worked to the disadvantage of pork producers.

The controversy had an unfortunate effect on the relationship between the principal protagonists. Wallace and Hoover developed strong antagonisms toward one another, antagonisms which continued after the termination of World War I. When Hoover's name was mentioned in 1920 as a possible presidential candidate, the farm editor launched a bitter campaign against him, rehashing all of the arguments against the Food Administration's handling of the hog matter and urging farmers to unite in opposition to any move to place Hoover in the Presidency.[48] "It would be a real misfortune," Wallace wrote to a friend, "if he [Hoover] should succeed in securing the nomination in either of the political parties." The neutral policy of *Wallaces' Farmer* notwithstanding, the editor felt obligated "to warn our readers against a possible candidate for nomination whom we regard as distinctly hostile to their interests." [49] Wallace's editorials aroused so much controversy that they provoked a discussion in Congress and forced Hoover to write a letter to the Senate in his own defense.[50] Less than a year later the two adversaries found themselves members of the same cabinet.

Henry C. Wallace had brought to the Department of Agriculture a wide and varied background. In many respects his experiences prior to entering the Harding cabinet served to prepare him for a multitude of problems he would meet in his new posi-

[48] *Wallaces' Farmer* 45 (February 13, 1920): 518–19 and 45 (March 19, 1920): 908.
[49] Wallace to H. S. Irwin, May 18, 1920; and Wallace to A. V. Mather, April 21, 1920, Wallace Papers, Box 2.
[50] *Congressional Record,* 66th Cong., 2nd Sess., pp. 3363–64, 3383.

tion. Some aspects of his background, on the other hand, were to hinder him in carrying out his responsibilities.

Certainly an important factor in Wallace's development was the influence of his father. Uncle Henry's approach to agricultural problems and the kind of solutions he conceived, his militancy and his irascible temperament, an intensity and determination of character, a kind of blind subjectivity toward certain questions, his political views and personal prejudices, his enjoyment of life and relaxed way among friends—all these were reflected in his son. The two men had a much closer relationship than usually found between father and son. Working together to build *Wallaces' Farmer,* they developed a strong and enduring mutual respect, while the close association on the paper tended to modify and blend their personalities and thinking and reduce their differences. That the father became a model which young Wallace, perhaps unconsciously, came to emulate is evident from his career as editor and later as Secretary of Agriculture. Symbolically, Uncle Henry's name remained as editor of *Wallaces' Farmer* for several years after his death.

Wallace's experience in practical farming also affected the way he performed and thought as Secretary of Agriculture. In the 1880's he learned firsthand what it was to be a farmer with a precarious dependence upon favorable economic and weather conditions. When farmers in the early 1920's complained of low prices, stringent credit, excessive service charges, droughts, and floods, Wallace had more than an academic understanding of their situation. Having experienced similar problems, he had an empathy with them and an appreciation of their plight which would have been impossible were it not for his own practical experience with farming. On occasions, however, his identification with the farming community was to prevent him from seeing the problems of agriculture within the broad framework of the entire economy and from taking a realistic approach in seeking solutions. A strong agrarian bias, ironically, would prove to be his greatest weakness as Secretary of Agriculture.

Although brief, Wallace's career as a college professor contributed to the determination of his attitudes. Interestingly, he never considered himself a member of the intellectual community and was, in fact, critical of many aspects of the academic approach to farming. To his thinking, professional agriculturalists failed to understand the basic problems of farmers because they remained sealed away in the nation's colleges and universities, oblivious to conditions as they actually existed among producers. This did not mean, certainly, that academicians could not be, or for that matter were not, of real service to producers. Enthusiastic about the scientific agricultural methods being developed in the colleges, Wallace, both as editor and Secretary of Agriculture, encouraged farmers to adopt them. But he believed that professionals could be of greater help if they would only become more familiar with the real world of farming and its problems, economic as well as technical. Shaped and reinforced at Iowa State Agricultural College, his attitudes toward agriculture as taught in the colleges would influence his approach to the postwar farm situation.

As secretary of the Corn Belt Meat Producers' Association, Wallace gained an appreciation of the power of collective action and learned the techniques of controlling and effectively using a farm pressure group. And his experience with the organization, as noted, also contributed to a growing awareness of the problems of agriculture and an increasing sophistication in meeting them. Fifteen years with the association provided Wallace with a valuable practical education, an education he was to put to good use as Secretary of Agriculture.

Perhaps most important in shaping his attitudes and approach was his association with *Wallaces' Farmer*. A long career in agricultural journalism gave him both an understanding of practical farming and insight into the economic, political, and social factors which largely determined the conditions under which producers had to operate. As an editor, he developed the techniques of arousing concern and guiding public opinion, invaluable assets to anyone in public office. In response to the postwar agricultural crisis, furthermore, he introduced in the columns of *Wallaces' Farmer* many of the measures he would later advance and support as Secretary of Agriculture. Since Wallace had sketched the outlines of a farm program before entering the Harding cabinet, he had the obvious advantage of a ready basis upon which to build. But there was also the disadvantage of a stubborn commitment to these earlier ideas which was to result in an unfortunate inflexibility in his new position.

Despite certain drawbacks, most of which would come to light later, Wallace appeared to be well prepared to take over the Agriculture Department. But technical qualifications for political appointments are rarely, if ever, prime considerations. And so it was with the selection of Harding's Secretary of Agriculture; yet another ingredient was necessary. In addition to being well-grounded in the field of agriculture, Wallace was also a prominent Republican, a factor which above all others accounted for his appointment.

Chapter III / CAMPAIGN AND APPOINTMENT

I T was natural that Wallace should have been a Republican. Born and reared in the heavily Republican Middle West among agricultural interests which generally supported the party, he turned naturally to the GOP. In a kind of negative way, a deep-seated aversion to radicalism was also a factor influencing his political preference, as the presence of an aggressive faction of former Populists in the Democratic party gave it a tinge of extremism which tended to alienate Wallace and reinforce his Republicanism. Most important in shaping his political views, however, was the close personal association with his father. Working with the dynamic Uncle Henry for much of his life, Wallace was certain to be influenced by the ideas and attitudes of the older man. His father's staunch support of the Republican party and the prominent position he held within its ranks were powerful determinants in the development of Wallace's thinking. Just as the elder John Wallace had instilled in Uncle Henry strong political convictions, so Uncle Henry through his own example unconsciously molded Henry C. into an avid Republican. Even when young Wallace and his father temporarily left the party to back Roosevelt in 1912, they were satisfied that their action was not one of political heresy. Rather they believed it to be in support of "true" Republicanism, as they understood it, and against the attempt of a minority faction to change party philosophy.

Although normally somewhat to the left of Republican leadership after 1909, Wallace played an active role in the party. As Uncle Henry had earlier, he became an influential member of the progressive faction of the Iowa organization. So prominent was his position, in fact, that some Iowa Republicans attempted to persuade him to run for the party's gubernatorial nomination in 1920. Having no political aspirations, however, he declined to enter the race. Consistent with family tradition, he declared a preference for working in the background and leaving the quest for political office to others.[1] Also like his father, Wallace gained prominence in the Republican national organization and earned a reputation among party members as an expert on farm policy. In 1918 Gifford Pinchot, who had risen to a high place in the party during the Theodore Roosevelt administration, asked him to become a member of the Republican agricultural advisory board. Though he turned down the position, the offer was indicative of Wallace's standing among national Republicans.[2]

Since Iowa was predominantly Republican, the party controlled most of the state public offices. As a ranking member of the party, Wallace was acquainted with many of these officeholders and enjoyed close friendships with some. Among his best friends was the state's leading Republican, Albert B. Cummins. The two men first met through Uncle Henry, who had worked with Cummins in the Farmers' Protective League, an organization formed in 1880 by the Agricultural Editors' Association of Iowa to fight the barbed wire monopoly in the state. Cummins, at the time a rising young Des Moines attorney, served as the or-

[1] Interview with Donald R. Murphy, Des Moines, Iowa, August 29, 1965.

[2] Wallace to Pinchot, August 12, 1918, Henry C. Wallace Papers, University of Iowa Library, Iowa City, Box 1.

ganization's legal counsel and eventually succeeded in bringing about the downfall of the monopoly.[3] Politically ambitious, he served several terms in the state legislature, captured the Iowa governorship in 1902, and finally won the Senate seat vacated by the death in 1908 of another prominent Republican, William Boyd Allison. Through his affiliation with the Republican party, Wallace developed deep respect and admiration for Cummins.

As was true of many state organizations at that time, the Iowa Republican party was plagued with factionalism. Groups of varying shades of progressivism and conservatism vied for control and fought over policy. One of the bitterest intraparty controversies of the organization's history broke out in 1920 when Cummins was up for re-election to the Senate. Having risen in the Republican party as leader of the progressive wing, he had earned a reputation as an ardent advocate of reform. In the postwar period, however, a former supporter, Smith W. Brookhart, challenged his position as head of the progressive faction, charging that Cummins had deserted his friends and joined the conservatives. The principal cause of Brookhart's disillusionment was the Transportation Act of 1920, which the Iowa senator had cosponsored with Representative John L. Esch of Wisconsin. The measure set down conditions for the return of the nation's railroads to private control after the war, during which they had been operated by the federal government. An outspoken advocate of government ownership of the roads, Brookhart insisted that despite provisions for increased federal regulation the terms of the act were highly favorable to railroad management, in particular the provision guaranteeing the lines a profit of five and one-half per cent on capital investment. With enactment of the bill in

[3] Henry Wallace, *Uncle Henry's Own Story of His Life* (Des Moines: Wallace Publishing Company, 1917–1919), vol. 3, p. 26; and William L. Bowers, "The Fruits of Iowa Progressivism, 1900–1915," Master's thesis, State College of Iowa, 1958, pp. 29–30.

February 1920, he announced his intention to run against Cummins in the Republican primary later that year.[4]

Wallace's position on the railroad question was clear. Although he had seriously considered the possibility of federal ownership earlier, the experience with government operation during the war had been enough to disabuse him of any such ideas. Wartime handling of the roads, according to the farm editor, was inefficient, inadequate, and too much involved with political considerations. "Theoretically, the government should own these great public utilities," he held, "[but] practically, we are as yet incapable of operating them efficiently." Thus Wallace was firmly set against all proposals for either government ownership or the retention of federal control. And while he recognized several weaknesses in the Esch-Cummins bill, he urged its adoption in view of President Wilson's announcement that the railroads would be returned to private operation in 1920, with or without legislation regulating the transfer. Wallace considered the increased power of the Interstate Commerce Commission provided for in the bill essential for the protection of agricultural interests against unwarranted rate increases after the return of the roads to private control.[5]

Apart from Wallace's personal regard for the Iowa senator and his attitude on the railroad question, there was yet another factor which led him to support Cummins in the Republican primary. Politically well to the left of Wallace, Brookhart was advancing a number of proposals, of which government ownership of the railroads was only one, that violated the farm editor's con-

[4] Jerry A. Neprash, *The Brookhart Campaigns in Iowa, 1920–1926* (New York: Columbia University Press, 1932), pp. 30–32; and Ray S. Johnston, "Smith Wildman Brookhart: Iowa's Last Populist," Master's thesis, State College of Iowa, 1964, pp. 82–84.

[5] *Wallaces' Farmer* 44 (April 18, 1919): 893; 44 (December 26, 1919): 2563; 45 (January 9, 1920): 79; and 45 (January 23, 1920): 258.

cept of the proper sphere of federal power. To be sure, Wallace considered himself a progressive, but he viewed Brookhart as too radical and too much inclined toward impractical, visionary schemes. The challenger's greatest support, furthermore, came from organized labor and the Farmers' Union, an organization representing the state's more militant farmers. And since in Wallace's opinion both of these groups favored extreme and irresponsible programs, their positions served to solidify his opposition to Brookhart. Interestingly, though not important in determining the farm editor's stand, James M. Pierce and the *Iowa Homestead* also backed Brookhart.[6]

Despite Cummins' wide popularity in Iowa, Brookhart's candidacy developed into a real threat to the incumbent senator. As postwar economic instability turned into full-scale depression in 1920, the challenger's campaign gained momentum. Particularly distressed were laborers and marginal farmers, the very groups to which Brookhart's candidacy appealed most. By May some ranking members of the Republican party felt that he had a good chance of winning the primary to be held the following month, and they warned Cummins, who had refused to take his opponent very seriously, to step up his campaign.[7]

Alarmed by the prospect of Brookhart representing Iowa in the United States Senate, Wallace departed from his journal's nonendorsement policy and openly supported Cummins in *Wallaces' Farmer*. The paper sent questionnaires to both candidates requesting their views on issues relating to the railroads, agricultural legislation, and political alliances between farm and labor

[6] Neprash, *The Brookhart Campaigns in Iowa,* p. 32; and Johnston, "Smith Wildman Brookhart: Iowa's Last Populist," p. 86.

[7] Wallace to Cummins, May 3, 1920; Louis C. Kurtz to Cummins, May 5, 1920; Cummins to Kurtz, May 6, 1920, Albert B. Cummins Papers, Iowa State Department of History and Archives, Des Moines, Box 20; and Johnston, "Smith Wildman Brookhart: Iowa's Last Populist," pp. 86–87.

groups. Printed in late May, Cummins' reply, with no accompanying editorial comments, appeared on the same page as a petition signed by 35 farmers endorsing his candidacy.[8] The following week, and only a few days before the election, Wallace published Brookhart's statement, but included parenthetical comments on the answers. "It is inconceivable," the editor concluded in his attack, "that the intelligent people of Iowa should consider for a moment depriving the state and the nation of the services of Senator Cummins . . . and send Colonel Brookhart, who displays such an abyssmal [sic] ignorance on railroad and financial matters." [9] Wallace also used his influence and position in the Corn Belt Meat Producers' Association to further Cummins' candidacy.[10]

Cummins won a narrow victory in the primary, an outcome which of course immensely pleased Wallace. At least one prominent Iowa Republican believed that the farm editor had played an important part in the results of the election. In a letter to Harding, the state's junior senator, William S. Kenyon, noted that Wallace "was of inestimable value in Cummins' campaign." [11] Complimenting Iowa farmers for "a mighty good piece of work" in renominating Cummins, Wallace pronounced the election a victory over "the forces of disorder and industrial socialism . . . and hair-brained [sic] schemes of government finance which if put in effect would certainly result in business catastrophe." [12] In the fall Cummins went on to win re-election to Congress. Undoubtedly, Wallace's strong support for him and his contributions to the primary campaign were factors in

[8] *Wallaces' Farmer* 45 (May 28, 1920): 1472.

[9] *Ibid.*, 45 (June 4, 1920): 1512.

[10] Wallace to Sykes, June 1, 1920; and telegram, Sykes to Wallace, June 2, 1920, Wallace Papers, Box 2.

[11] Kenyon to Harding, July 12, 1920, Warren G. Harding Papers, State Historical Society of Ohio, Columbus, Box 533.

[12] *Wallaces' Farmer* 45 (June 18, 1920): 1602.

the senator's early endorsement of his appointment as Secretary of Agriculture.

While Wallace was from the outset unwavering in his endorsement of Cummins in the Iowa primary, he was "not very enthusiastic over any particular candidate" for the Republican presidential nomination in 1920. The death of Theodore Roosevelt the year before removed the one man who would have received his unqualified support. Senators Hiram W. Johnson and Robert M. La Follette, both frequently mentioned possibilities, had recommended measures which Wallace regarded as dangerously radical. "Too extreme in his views to make a satisfactory presidential candidate" was the editor's opinion of Johnson. As for Hoover's candidacy, he had already made his feelings clear. In a letter to a Nebraska Republican, he urged him "to get the political leaders to understand that the farmers in your state have no use for Hoover." As early as April 1920 Wallace had decided that Warren G. Harding had a good chance of securing the nomination. Although either the Ohio senator or Hoover "would be quite satisfactory to Big Business," he believed "their first effort will be made to name Harding." Despite, or perhaps because of, the fact that Wallace had never met Harding and knew little about him, neither was he pleased with this possibility.[13]

Of the two leading Republican contenders, Frank O. Lowden and General Leonard Wood, apparently either would have been satisfactory to Wallace. He believed that Lowden had established a creditable record as governor of Illinois and had demonstrated an understanding of farm problems. Former Army Chief of Staff, a moderate progressive, and Roosevelt's heir apparent in the party, Wood represented better than any of the major candidates Wallace's political views. The General, he thought, "would make a most admirable president. He is not an extremely military man,

[13] Wallace to A. V. Mather, April 21, 1920; Wallace to F. C. Crocker, April 13, 1920, Wallace Papers, Box 2; and Murphy Interview.

as some people would like to have us believe. He is a very level-headed, conscientious, straight forward man with a splendid record as an administrator. . . ." Wallace's main reservation regarding both Lowden and Wood was his fear that politically they would be weak candidates.[14]

Meeting in Chicago in June 1920, the Republican national convention bore out Wallace's prediction of the previous April in selecting Harding as its presidential candidate and Massachusetts governor Calvin Coolidge as his running mate. While the party was busy with the balloting, Wallace was assisting the platform committee in the formulation of agricultural policy. The Republican farm plank of 1920, containing as it did many measures which *Wallaces' Farmer* had been advancing for some time, clearly reflected his influence. It upheld the right of farmers to organize marketing associations free from legal restraint; recognized the need for a government program of agricultural economics, better rural transportation systems, and improved rural credit facilities; pledged encouragement of the agricultural export trade and an end to federal price-fixing; and promised adequate farm representation on government agencies affecting agricultural interests. Although it offered little that was new, the plank pleased farmers and would be instrumental in securing the support of the large majority of producers outside of the Democratic South. Expressing the opinion of many Republican candidates in farm states, Cummins pronounced himself "entirely satisfied with it." [15]

Soon after the convention Harding and Harry M. Daugherty began formulating plans for the approaching campaign. In the

[14] Wallace to A. V. Mather, April 21, 1920, Wallace Papers, Box 2; and interview with James W. Wallace, Des Moines, Iowa, December 21, 1966. Wallace, however, opposed Wood's appointment as Secretary of War in the Harding cabinet. Wallace to Harding, February 10, 1921, Harding Papers, Box 655.

[15] Cummins to Wallace, July 15, 1920, Cummins Papers, Box 20.

previous two presidential elections the Democratic party had shown surprising strength in the traditionally Republican farm states of the Middle West. For the purpose of developing strategy to regain the party's lost position in this area, Daugherty recommended that Harding call together a group of farm experts to counsel him on agricultural policy. Such a meeting, he pointed out, would have the added advantage of demonstrating that "agricultural interests will be consulted and advised in regard to all matters pertaining to agriculture." While Daugherty left the selection of the group up to Harding, he thought that it would be well to invite Wallace. In addition, he suggested that Harding should consult Iowa's junior senator, William S. Kenyon, who was an aggressive representative of agricultural interests, well respected among the farmers of his state, and a recognized agrarian spokesman in Congress.[16]

Following Daugherty's suggestion, Harding wrote to Kenyon soliciting his advice. Recognizing "the importance of a thoroughly sincere and becoming appeal to the farming interests of this country," the Ohio senator went on to assure his colleague that his concern for agriculture stemmed not only from political considerations but also from a belief that the improvement of farm conditions was "essential to the welfare of our common country." In wholehearted agreement, Kenyon insisted that the Republican party had to develop an imaginative approach to agricultural problems. "The best man I know of in the whole United States, with reference to this particular subject, is Mr. Henry Wallace," he advised Harding. "Wallace is a sturdy Scotchman, staunch, level-headed, and knows agricultural problems." Furthermore, according to Kenyon, *Wallaces' Farmer* "is the gospel to the farmers." Recommending that Harding invite Wallace to his home in Marion, Ohio, to discuss agricultural

[16] Undated note from Daugherty to [?, probably George B. Christian, Jr., Harding's secretary], Harding Papers, Box 533.

questions, Kenyon assured the Republican candidate that "this is the best practical suggestion I can make." [17]

Although Harding decided against calling a meeting of agricultural experts, he saw fit to invite Wallace to Marion. Until their first campaign conference on July 26 the two men barely knew each other, but this meeting marked the beginning of a close professional relationship and a cordial personal friendship. After the discussion Wallace told newspaper reporters that he was favorably impressed and pleased to find that Harding had "the vision of what must be done and . . . the courage to undertake and insist upon it." In a report to Cummins, Wallace indicated that he was most satisfied with the visit and that his enthusiasm for the Ohio senator had greatly increased.[18] Also favorably impressed, Harding asked Wallace to act as his campaign advisor on agricultural questions, a task which the farm editor accepted and undertook with characteristic zeal.

Wallace's work in the Harding campaign demonstrated both a sensitivity to agricultural problems and a political sagacity. Despite the fact that the Republican candidate had slight grasp of the crisis which was descending upon farmers in the postwar period, Wallace's efforts to educate him were at least partially successful. While he supplied Harding with a vast amount of material and practical suggestions to direct his statements on agriculture, the farm editor recommended that the best guide Harding could use was the agricultural plank of the party's platform.[19] During the course of the campaign, Wallace also had occasion to use his contacts among farm interests to good

[17] Harding to Kenyon, July 7, 1920; and Kenyon to Harding, July 12, 1920, *ibid.*

[18] Cummins to Harding, July 27, 1920, *ibid.,* Box 694 and Cummins Papers, Box 20; Wallace to Harding, August 1, 1920, Harding Papers, Box 533; and *New York Times,* July 27, 1920, p. 3.

[19] Wallace to Harding, August 2, 1920; and Wallace to Harding, August 18, 1920, Harding Papers, Boxes 533, 525.

advantage. In one interview, for instance, Harding made an innocent misstatement concerning the policy of the American Farm Bureau Federation, to which the federation took exception. Realizing that it would be unwise to alienate an organization of the power and influence of the Farm Bureau, Wallace met with his friend James R. Howard, president of the federation, and quickly cleared up the misunderstanding.[20] Further demonstration of his political awareness came when he advised Harding not to accept an invitation to speak before the National Board of Farm Organizations and, in fact, counseled against addressing any single farm group. "In this way," he ventured, "there would be no cause for jealousy between the different organizations." [21]

Departing once again from its "non-partisan" editorial policy, *Wallaces' Farmer* gave solid support to the Republican party in 1920. Though the journal avoided direct endorsement of Harding, it was encouraged that "one of the great political parties has made definite pledges for certain definite things in which the farmer is much interested," leaving no doubt, of course, that the Republican party was the one in question. The agricultural plank of the Democratic platform, on the other hand, was "devoted to asserting claims for legislation enacted some years ago." Finding the plank "backward . . . [and] evidently drawn up by politicians who had no knowledge of the needs of agriculture and were not willing to listen very patiently to those who did," *Wallaces' Farmer* was certain that producers would have little trouble determining which organization had their interests at heart.[22]

Wallace's principal piece of work in the 1920 campaign was on Harding's major agricultural address, delivered in early Sep-

[20] Wallace to Harding, September 3, 1920; J. R. Howard to Harding, September 8, 1920; and Wallace to Harding, September 10, 1920, *ibid.*
[21] Wallace to Victor Heintz, July 30, 1920, *ibid.*, Box 533.
[22] *Wallaces' Farmer* 45 (June 18, 1920): 1602 and 45 (July 9, 1920): 1714.

tember at the Minnesota State Fair. Although he did not actually write the speech, as one Iowa newspaper claimed after the election, Harding followed his suggestions closely.[23] Along with specific recommendations on what should be included, Wallace advised that the speech must be written "in such a way as to convince your hearers and those who read it afterward, that you have a thorough grasp on the agricultural situation in all of its bigness. . . ."[24] Alarmed over newspaper statements that the Minnesota address was to be devoted to the question of American entry into the League of Nations in addition to farm policy, the farm editor objected to Harding's secretary that if this were the plan he considered it poor strategy. If both subjects were dealt with in the same speech, Wallace feared that the league pronouncements would monopolize press accounts and draw attention away from the statements on agriculture. "As I see it," he cautioned, "[the] speech on agriculture ought to be given absolute right of way . . . and nothing should be said which is likely to interfere with this."[25] Apparently Wallace's view prevailed, for the speech dealt solely with farm questions.

Harding's address at the Minnesota State Fair was primarily a restatement of the Republican agricultural plank. He reiterated and elaborated upon party pronouncements concerning agricultural economics, rural credit, and price-fixing. In addition, however, he clarified the Republican position on an issue which was to become of considerable importance in the 1920's—the agricultural tariff. While the party platform made vague reference to equal protection, it did not deal directly with the subject as it applied to farm products. But Harding left no doubt at Minneapolis that, in the event of his election, the tariff would be a major

[23] *Fort Dodge Messenger,* January 11, 1921, p. 3.
[24] Wallace to Harding, August 18, 1920; and Harding to Wallace, August 30, 1920, Harding Papers, Box 525.
[25] Wallace to George B. Christian, Jr., August 20, 1920, *ibid.,* Box 533.

weapon against agricultural problems. "If we are to build up a self-sustaining agriculture . . . ," he asserted, "the farmer must be protected from unfair competition from those countries where agriculture is still being exploited and where the standards of living on the farm are much lower." Hinting at the need for a full-fledged federal agricultural policy, he declared that the government must give greater attention to farm problems and work to build a sound system in which producers were guaranteed a fair return on their labor and investments.[26]

As laid out in the Minnesota speech, the Republican agricultural program suggested a more comprehensive approach than farmers had come to expect from either of the two major parties. Although it had obvious flaws and would prove inadequate as the farm crisis deepened, the program nevertheless indicated an awareness on the part of the Republican party of the need for some kind of action by the federal government. Representing the opinion of many producers, the *Prairie Farmer* pronounced Harding's speech "remarkable because it was the first time a presidential candidate has ever recognized the importance of agriculture by devoting an entire address to it." [27] The Democratic party, by contrast, seemed little interested in appealing to the rural voter and content with recalling what the Wilson administration had done for the farmer.[28]

Wallace was largely responsible for the Republican party's unusually close attention to agricultural matters and its attempt to attract the farm vote in the 1920 campaign. Not only was he instrumental in the formulation of policy, but he took an active part in getting information into the hands of the farmers. He had

[26] Speech reprinted in *Wallaces' Farmer* 45 (September 17, 1920): 2171.

[27] *Prairie Farmer,* September 18, 1920, p. 20.

[28] Henry L. Rofinot, "Normalcy and the Farmer: Agricultural Policy under Harding and Coolidge, 1920–1928," Doctoral dissertation, Columbia University, 1958, pp. 10–12.

Harding's Minnesota speech printed in pamphlet form and distributed throughout agricultural areas. "It is going to make an excellent campaign document in those closer states where the farm vote will be the determining factor," he advised Harding. In a flier entitled "Heart to Heart Talk to Farmers," Wallace himself carefully explained the Republican position on various agricultural questions, and in a weekly newsletter distributed to some 3,500 newspapers he delivered party statements on the farm situation. Working with Republican organizations in several states, the farm editor also endeavored to develop, as he wrote to his brother, a "systematic campaign for the farmer vote in the principal doubtful states." [29]

Under Wallace's direction the Chicago campaign headquarters prepared a full-page advertisement which appeared in a number of agricultural journals. Topped by pictures of Harding and Coolidge and under the caption "The Republican Party and the Farmer," the page was devoted mainly to attacks on the Democratic administration and its "unwise, unsympathetic policy." To mention only one of the things Wallace chose to focus upon, he asserted that price-fixing had limited farm income during the war, while there was no corresponding limit placed on what producers might be charged for the things they had to buy. As a result, charged the advertisement, farmers had been exploited by profiteers and denied their just share of the national income. The Republican party, on the other hand, "believes that the farmer, whose industry is the very foundation of our national prosperity, should have his fair share of the wealth which his labor and enterprise creates." As the advertisement illustrated, Wallace's principal campaign strategy was "continual emphasis on the incompetent meddling of the Democratic administration." [30]

[29] Wallace to Harding, September 10, 1920; Wallace to Harding, September 2, 1920, Harding Papers, Boxes 535, 525; and Wallace to Dan A. Wallace, August 16, 1920, Wallace Papers, Box 2.
[30] See, for instance, *Prairie Farmer*, September 18, 1920, p. 17; and Wallace to Harding, September 14, 1920, Harding Papers, Box 533.

Harding won an impressive victory over Ohio Governor James M. Cox, his Democratic opponent in the November presidential election, losing only 11 states, all in the heavily Democratic South. To hold that the Republican stand on agriculture was a major factor in the outcome of the election would be a distortion. There were more important issues, such as the League question, labor unrest, and prohibition. General disillusionment after eight years of Democratic control, furthermore, probably had more to do with Harding's victory than anything else. But Republican agricultural pronouncements, as noted, won the support of many farmers and accounted in large part for Harding's strong showing in the Middle West and the Great Plains. In the election of 1916 the Republican party carried only seven of twenty states in the Middle and Far West, while in 1920 it won all of them. The proportion of Harding's popular vote exceeded that of Charles Evans Hughes, the Republican presidential candidate in 1916, by 20 per cent or more in 12 of those states. Harding's greatest margin over Cox, moreover, came in heavily agricultural North Dakota where he captured 78 per cent of the popular vote. To Wallace, most Republicans agreed, went much of the credit for the party's recovery in the farming areas of the West.

After the election President-elect Harding began work on the formation of his cabinet. Despite a good deal of political maneuvering within the Republican party itself, there was surprisingly little public interest in Harding's selections for the various posts. Debate on who should receive the portfolio of agriculture was rather lively, however, and at least in some circles generated considerable controversy.

As early as July 1920 Wallace was mentioned as the probable candidate for Secretary of Agriculture if Harding won in the presidential race. Early in his campaign the Republican standard-bearer had promised to name a real "dirt" farmer to the position in the event of his election, and shortly after this announcement he called Wallace to Marion for their first meeting. Since the editor had indeed been a farmer, his appointment as campaign advisor appeared to be a good sign that he would be Harding's choice for the agriculture post. With the Republican candidate's victory in November, speculation on members of the new cabinet invariably conceded that Wallace would most likely head the Department of Agriculture.[31] "With a Republican administration, Harding at the helm," wrote a friend to Wallace the day of the election, "I take it that you will be the next Secretary of Agriculture." In view of their close association during the campaign, Wallace's hometown newspaper observed that "it is likely that Harding in his own mind picked Wallace earlier and more definitely than any other member of his cabinet." [32]

No doubt these speculations were correct. Three days before the election Harding wrote to Wallace to thank him for his assistance during the campaign and "to say to you again that if the verdict on Tuesday is what we are expecting it to be, I shall very much want your assistance in making good the promises which we have made to the American people." [33] At sometime during the campaign, then, it would appear that the two men had discussed a position for the farm editor in the administration. Whether Harding had specifically asked him to become Secretary of Agriculture in the event of a Republican victory is, of course, conjecture. But he clearly would have had to be considering some position in the Agriculture Department, and for him to have offered Wallace anything less than the secretaryship seems most unlikely.

[31] *New York Times,* July 27, 1920, p. 3. Wallace's files contain hundreds of newspaper clippings from all over the country pertaining to the selection of the cabinet. Although other names were given as possible choices, Wallace was considered in every case to be Harding's first pick for Secretary of Agriculture. Wallace Papers, Box 5.

[32] John M. Evvard to Wallace, November 4, 1921, *ibid.,* Box 3; and *Des Moines Register,* January 11, 1921, p. 1.

[33] Harding to Wallace, November 1, 1920, Harding Papers, Box 533.

Wallace met with Harding only once between the election and the inauguration, and that was when he traveled to Marion in late December 1920. At this meeting Harding's formal offer and the editor's acceptance of the agriculture appointment probably occurred. At any rate, Wallace knew at least by the beginning of the next year that he was to be the department's next head, for he alluded to it in a letter of January 3.[34] But the President-elect was not to announce the selection of his cabinet until February 25, and in the meantime there was a good deal of political maneuvering and discussion regarding the appointment of the next Secretary of Agriculture.

Agricultural and associated interests widely supported Wallace's selection as Secretary of Agriculture. Of the country's three major farm organizations, two, the American Farm Bureau Federation and the National Grange, backed him for the post. He also received direct endorsement from many of the state divisions of both organizations.[35] Though Wallace's opposition to Brookhart was no doubt part of the reason for the failure of the Farmers' Union to back him, more important was a fundamental disagreement between the editor and the organization on how best to attack agricultural problems. In addition, several of the country's leading agricultural journals, notably the *Prairie Farmer* and the *Farm Journal,* spoke favorably of Wallace's possible appointment.[36] Endorsement likewise came from numerous

[34] *New York Times,* December 12, 1920, p. 3; Wallace to A. S. Alexander, January 3, 1921, Wallace Papers, Box 3; see also Wallace to E. T. Meredith, January 4, 1921, Edwin T. Meredith Papers, University of Iowa Library, Iowa City, Box 6.

[35] Telegram, J. R. Howard and J. W. Coverdale [president and secretary of American Farm Bureau Federation] to Wallace, December 28, 1920; T. C. Atkeson [president of National Grange] to Wallace, January 1, 1921; L. H. Goddard to Wallace, February 9, 1921, Wallace Papers, Box 3; and *Des Moines Register,* January 7, 1921, p. 1.

[36] Charles F. Jenkins [editor of the *Farm Journal*] to Wallace, December 28, 1920, Wallace Papers, Box 3; *Prairie Farmer,* January 16, 1921, p. 4 and March 19, 1921, p. 8.

business interests connected with farming, in particular the implement industry.[37]

Iowa's two senators, as might well be expected, actively supported Wallace's assignment to the Agriculture Department. In a private conference with Harding in early December, Cummins reportedly pressed for the farm editor's appointment. A short time later he wrote to Wallace expressing the hope that "what we are trying to do here [in Washington] will come out all right and that before long you will be a resident of Washington. . . ."[38] Kenyon, who actually took a more active interest in farm matters than Cummins, concurred in his colleague's feelings on the subject. "I believe there is no step," he announced to the press shortly after Harding's election, "that could be taken just at this time that would be so beneficial to the country as the selection of Henry Wallace. . . . He knows the things the farmers need and they have absolute confidence in him." Kenyon also took up the matter personally with Harding.[39]

Incumbent Secretary of Agriculture Edwin T. Meredith, a fellow Iowan and owner of *Successful Farming,* publicly praised Wallace as the best choice the Republicans could make to head the department. In a premature letter in December 1920, he offered his congratulations on Wallace's appointment and his aid in getting him established in the new position the following March. After the Harding administration took over, Meredith

[37] Henry J. Barbour [Avery Company] to Wallace, December 28, 1920; Keller J. Bell [E. W. Ross Company] to Wallace, January 5, 1921, Wallace Papers, Box 3. Wallace's files contain hundreds of endorsements from friends and people associated with agriculture. The Harding papers also contain similar correspondence as well as letters of opposition. Harding Papers, Box 389.

[38] *Des Moines Register,* December 7, 1920, p. 4; and Cummins to Wallace, December 29, 1920, Wallace Papers, Box 3. By this date, Wallace, as noted, had probably already been offered and had accepted the position. Why Cummins had not been advised, if this were the case, is not clear.

[39] *Des Moines Register,* November 14, 1920, p. 1; and *Fort Dodge Messenger,* December 28, 1920, p. 1.

wrote in his own journal that "Mr. Wallace is a clear thinker, a wise counselor, a good fighter for what he considers a just cause and a man who can be depended upon to study agricultural problems from the broad viewpoint of the greatest good to the greatest number." [40] Henry C. Taylor, a member of the Agriculture Department and soon to become one of the new secretary's closest friends and most ardent supporters, wrote to Wallace shortly before the election urging him to accept the headship if, as he anticipated, Harding were to tender it. He assured the farm editor that his appointment was something to "which the men of the Department would look forward with much satisfaction." [41]

In light of their controversy during World War I, Hoover's support of Wallace seemed improbable. Many years later Hoover was to remember that Harding had asked his opinion on the selection of the farm editor and that he had given his endorsement.[42] The evidence clearly suggests, however, that it was Hoover rather than the President-elect who took the initiative in the matter. When it was obvious that Wallace was a near certainty for the portfolio of agriculture, Hoover wired Harding that he approved of the appointment. Feeling that Wallace was "admirably fitted for the work" and that he "would render a great sense of teamwork in the real interest of agriculture," he judged the farm editor a wise choice for the post. Because of their past relations, Hoover was confident that his support of Wallace would be the more important. There was no indication that he had previously discussed the subject with Harding.[43]

[40] E. T. Meredith to Wallace, December 31, 1920, Wallace Papers, Box 3 and Meredith Papers, Box 6; see also Meredith to Wallace, March 3, 1921, *ibid.; Des Moines Register,* January 17, 1921, p. 1; and *Successful Farming,* April, 1921, p. 6.

[41] Taylor to Wallace, October 28, 1920, Wallace Papers, Box 3.

[42] Herbert C. Hoover, *The Memoirs of Herbert Hoover* (New York: Macmillan, 1952), vol. 2, p. 109.

[43] Telegram, Hoover to Harding, January 12, 1921, Harding Papers, Box 389. Wallace wrote to Hoover thanking him for his support. Wallace

Although Wallace's selection as Secretary of Agriculture was gaining wide support, there were centers of strong opposition. Most vocal in denouncing his possible appointment was the meat packing industry. Wallace's endorsement of the Kenyon-Kendrick bill, a measure providing for strict federal regulation of packers and livestock yards, had brought down upon him the wrath of the industry. Representing the Corn Belt Meat Producers' Association, he had petitioned Congress in support of the legislation, and as editor of *Wallaces' Farmer* he had endeavored to stimulate interest in the bill in an unsuccessful effort to bring about its passage. Intent on blocking Wallace's appointment, the big meat packers alerted friends in government and put their powerful lobby to work, but their efforts proved self-defeating. "The fact that the organized Meat Industry . . . was against your appointment," wrote a friend to Wallace, "makes it all the more agreeable to the American farmer." [44] The farm editor's hometown newspaper surmised, and probably correctly, that for Harding to yield to this pressure would cost him the support of a large part of the farming community. "Senator Harding," the paper thought, "is too good a politician to invite just that sort of trouble." By January 1921 the *Washington Post* had decided that the opposition of the packing interests had "so strengthened [Wallace] that he has become one of the absolute certainties of the cabinet." [45]

Wallace's interest in the Kenyon-Kendrick bill also drew the opposition of the livestock exchanges and marketing interests. "On his past performances," wrote a spokesman for the exchanges, "Henry Wallace is the last man who would have been acceptable to the livestock trade for the agricultural port-

to Hoover, January 18, 1921, Herbert C. Hoover Papers, Herbert C. Hoover Presidential Library, West Branch, Iowa, AK I-7.

[44] [?] to Wallace, January 3, 1921, Wallace Papers, Box 3.

[45] *Des Moines Register,* January 10, 1921, p. 4; and *Washington Post,* January 16, 1921, p. 11.

folio." [46] Personally calling upon Harding in late December, representatives of the marketing interests urged him not to place Wallace at the head of the department. Their efforts, however, had largely the same reaction as the packer opposition. "This sort of thing," Wallace observed, "has stirred up a lot of men who are friendly to me and who felt they should come to my rescue." [47] Although not so well-organized as either the packers or the exchanges, the country's lumbering interests made a weak attempt to influence Harding's decision on the agriculture portfolio. Because of Wallace's reputation as a staunch conservationist, lumbermen feared curtailment of cutting rights in the national forests, which were under the control of the Agriculture Department, if he were to become secretary. Amounting to nothing more than the issuing of a few press releases, their activities figured little in the cabinet debate. [48]

Although Wallace was the most frequently mentioned choice for the agriculture post, other names appeared as possible candidates. These included his former journalism partner at Ames, Charles F. Curtiss, Frank O. Lowden, Arthur Capper, and even Gifford Pinchot. Yet, despite the agitation against Wallace and in behalf of other candidates, Harding seems to have seriously considered only Wallace for the position. And it is likely that he had settled fairly definitely on him before the election.

Between Harding's November election and his official announcement of the cabinet selections on February 25 was an uncomfortable period for Wallace. Press reports regarding his possible appointment prompted hundreds of supporters to write to him asking what they could do to further his candidacy. Irritating to him because they seemed to imply that he was actively seeking the position, he answered all of these inquiries politely,

[46] *Sioux City Live Stock Record,* December 28, 1920, p. 1.
[47] Wallace to J. H. Mercer, January 13, 1921, Wallace Papers, Box 3.
[48] *Fort Dodge Messenger,* December 28, 1920, p. 1.

but firmly. "There is nothing . . . ," was his reply to one, "that I wish any of my friends to do for me in connection with a cabinet post. I am not now and never have been a candidate for it, nor have I any desire for it." To conduct a campaign for the purpose of securing a cabinet position, Wallace thought, was in "rotten bad taste." Despite the absence of official confirmation and his contention that "the talk in the newspapers is 'hot air,'" many people refused to believe that he had not already received the appointment and prematurely congratulated him, a situation which caused the farm editor considerable embarrassment. "The only thing I can do," Wallace confided to his brother, "is to grin a sickly grin and talk about something else." [49]

Wallace claimed that he was not anxious to become Secretary of Agriculture and accepted the position with considerable reservation. Comfortably settled in Des Moines, he faced with regret "the thought of being torn up by the roots and transplanted. . . ." [50] With a large financial investment in *Wallaces' Farmer,* furthermore, Wallace was understandably reluctant to leave, the more so because he did not have complete confidence in the business judgment of his brother John, under whose management the paper would be left. Knowing that Henry Agard would assume active editorship, however, helped to relieve his concern, for the proud father had great faith in the ability of his oldest son, maintaining that he had the finest mind of all the Wallaces in the previous 200 years. Family tradition also figured in Wallace's decision. "Justice-of-the-peace is the highest office ever held by any of our family," Uncle Henry enjoyed relating. "I

[49] Wallace to Edwin Dukes, February 8, 1921; Wallace to Dan A. Wallace, January 10, 1921; Wallace to J. B. Weems, November 9, 1920; and Wallace to Dan A. Wallace, January 19, 1921, Wallace Papers, Boxes 2, 3.
[50] Wallace to A. S. Alexander, January 3, 1921, *ibid.,* Box 3; and Pinchot's introduction to Henry C. Wallace, *Our Debt and Duty to the Farmer* (New York: Century, 1925), pp. vi–vii.

hope that no Wallace ever attains a higher office or aspires to it."
Given a family as imbued with the idea of tradition and continu-
ity as the Wallaces, this was a consideration which the new secretary had to weigh before making up his mind.[51]

Wallace's reason for accepting the appointment was simply
that he believed he could be of greater service to the country's
farmers in the Agriculture Department than publishing an agricultural journal. Intimately involved with agriculture for some
time, he had conceived the outlines of a federal program designed to solve the problems of farmers, and he was confident
that if his ideas were given a chance, it would mean at least a degree of relief for producers. In this many of his friends concurred. Urging Wallace to accept the cabinet position, Gifford
Pinchot assured him that therein "lay the line of greatest
usefulness." [52] In a letter to a friend, the farm editor admitted
that "I go with a great deal of reluctance and only because so
many people seem to think that for a time at least I can be of
more use to the farmers' interests of the country back there than
I can here." What was more, Wallace was satisfied that the incoming President would be receptive to his ideas and willing to
support a broad attack on farm problems. "I go to Washington,"
he told readers in his final editorial in *Wallaces' Farmer*, "with
less reluctance because from the beginning President Harding
has taken an advanced stand for a sound national policy as it related to agriculture. . . ." [53]

Harding's choice of Wallace for the Department of Agriculture was politically sound. A prominent Republican from an important agricultural state and editor of the highly respected

[51] Murphy Interview; *Wallaces' Farmer* 46 (March 11, 1921): 392;
and *Prairie Farmer*, March 5, 1921, p. 3.

[52] Pinchot's introduction to Wallace, *Our Debt and Duty to the Farmer*,
p. vii.

[53] Wallace to A. Sykes, February 28, 1921, Wallace Papers, Box 3;
and *Wallaces' Farmer* 46 (March 11, 1921): 392.

Wallaces' Farmer, Wallace had won the confidence of the country's farming community. He had come to be recognized as an expert on the problems of producers not only in the Republican party but throughout the Middle West. Beyond this his support of Roosevelt, close association with Cummins, and advocacy of agricultural reform legislation had earned for him a progressive label. Thus the liberal wing of the party strongly supported his appointment, the more so because of the dominant position of the old guard within the new administration, while party conservatives found him acceptable. That vested interests represented his major opposition, as noted, rendered his candidacy simply that much more attractive. Satisfactory to all factions within the party and appealing to the interests he would be serving, Wallace represented an ideal political appointment.

In addition to his political and technical qualifications for the post of Secretary of Agriculture, Wallace had gained Harding's trust and confidence during the campaign. Ill-prepared to speak on agricultural matters, the Republican candidate accepted Wallace's suggestions almost without reservation. When his pronouncements were well received among farmers, Harding developed a great respect for the farm editor's judgment and felt indebted to him for his role in the successful campaign. Working closely together to win the election, furthermore, the two men became personal friends. To the genial Harding support and friendship were things to be rewarded, a consideration which also must have contributed to his decision to make Wallace his Secretary of Agriculture.

Knowledgeable on the subject of agriculture, from a farming area, acceptable to agricultural interests, reasonably well known and respected, a loyal and prominent Republican, and a friend of the President-elect—few in the cabinet could boast as many qualifications for the posts to which they were assigned.

Chapter IV / THE FARM CRISIS
AND A PLAN OF ATTACK

As the new cabinet gathered on March 8, 1921, Wallace was surely contemplating the crisis which had recently descended upon the country's farmers. So severe had the situation become that he must have wondered whether the agricultural program he had worked out in his own mind and presented in *Wallaces' Farmer* would really be adequate even if it were adopted. Farmers, to be sure, had suffered economic depressions before and had endured until the return of better times. The new secretary could clearly remember the last one which began with the panic of 1893, and he had heard his father talk about the crises of the 1850's and 1870's. But the postwar depression seemed worse for farmers than earlier ones had been, perhaps because it followed a period of unprecedented farm prosperity, perhaps because it exposed as no previous one had the fundamental weaknesses of American agriculture.

Ever since pre-Civil War years the farming community had been declining economically in relation to the country's business and financial groups. Gradually replaced by expanding industry, agriculture in the latter nineteenth century finally lost its traditional place as the mainstay of the American economy. As evidence of this, farmers learned in 1890 that for the first time annual production of the nation's factories had surpassed their own in value. Although agricultural commodities continued to dominate the export trade, they declined from 82 per cent of the total in 1860 to 51 per cent in 1910. In yet another and in many respects more poignant way, the steady flow of population from rural to urban areas reflected the trend of the economy and the deteriorating position of the farm sector. Predominantly agrarian in composition throughout the nineteenth century, the United States was none the less gradually becoming a nation of city dwellers, a movement which carried over into the next century until in 1920 the rural population (communities of below 2,500) made up less than half of the country's total, with the farm population representing less than 30 per cent.[1]

Despite its increasingly subordinate position, agriculture enjoyed one of its most prosperous periods of American history in the first two decades of the twentieth century. Beginning in 1897 the country witnessed the convergence of several factors which redounded to the benefit of farmers. A population growing as a result of natural increase and mounting immigration secured the power to purchase farm products from the nation's rapidly developing industrial machine. What the domestic market could not absorb was channeled off through the export trade. Due to the disappearance of unoccupied land suitable for profitable farming, agricultural output failed to keep pace with the expansion in demand, and the problem of price-depressing surpluses which had plagued the farm sector in the latter nineteenth century disappeared. Responding to these favorable conditions, the real income of farmers rose at an unusually rapid rate. In the first decade of the new century farm prices increased 47 per cent while those of nonfarm products moved up only 18 per cent. The

[1] For the pre-World War I agricultural conditions see Gilbert C. Fite, *George N. Peek and the Fight for Farm Parity* (Norman: University of Oklahoma Press, 1954), pp. 5–8; and Edwin R. A. Seligman, *The Economics of Farm Relief* (New York: Columbia University Press, 1929), p. 14–15.

value of farm land appreciated faster than the price level was rising, moreover, giving producers an indirect profit on their capital investment. So favorable was the period 1900–1914, in fact, that farmers would later look back to it as the "golden age" of agriculture and seek to recover the prosperity it represented.

World War I climaxed agriculture's golden age by pushing farm incomes to still higher levels. With relatively little change in output, the marked increase in demand for food and fiber brought on by the war resulted in sudden price rises. At first a boon to farmers, the upheaval in Europe was ultimately a disruptive influence in the development of American agriculture which contributed greatly to the problems of producers in the 1920's.

Mounting demand for the products of American farmers came largely from abroad. As war spread across Europe, the established sources of food and fiber on the Continent were either overrun or unable to satisfy growing needs, thus forcing the Allied Countries to turn to the United States. The declining importance of the export trade in farm commodities was therefore temporarily reversed, as American agriculture responded to changing conditions on the other side of the Atlantic. Although welcomed by farmers at the time, this increased European demand augured ill, for it brought with it a renewed dependence on the foreign market.[2]

The entry of the United States into the war only pushed prices higher. Now farmers were called upon not only to supply the needs of the Allied Countries but to sustain the American war effort as well. The federal government, through the Food Administration, launched various programs to induce producers of food and fiber to increase output, and many of these included supporting market prices.

Eager to add to their holdings to take advantage of high

[2] *Agricultural Situation* (USDA), August, 1921, pp. 4–6; and Seligman, *The Economics of Farm Relief*, p. 16.

prices, farmers began buying land at a much more rapid rate than had prevailed during peacetime. The sudden change in demand resulted in a disproportionate rise in the value of farm land, particularly in the more favorable agricultural regions. Also with an eye toward increasing production, farmers mechanized their operations at an accelerated pace. To purchase land at inflated prices and acquire machinery many producers unwisely contracted sizable loans. Their willingness to go into debt was encouraged not only by good times but also by readily available money. In the midst of agricultural prosperity, rural banks were eager to make loans to farmers and in many cases actually encouraged them to borrow. The Federal Reserve Board, furthermore, in the interest of facilitating the financing of the war effort followed an easy credit policy.[3]

With the end of the war most farmers expected prices to fall, although few anticipated the drastic readjustment which was in store for them. The decline did not directly occur, however, as agricultural prices remained at artificially high wartime levels for a short while after the armistice. Several factors contributed to this situation. In the immediate postwar period the United States continued to extend credit to the bankrupt Allies, enabling them to maintain their purchases of American food and fiber and thus to sustain the foreign market upon which farmers had come to depend. For several months following the cessation of hostilities the Federal Reserve Board chose to continue its easy credit policy and did little to check the inflationary forces which had been set in motion during the war. Returning soldiers and the stimulus of wartime prosperity, furthermore, tended to strengthen domestic demand for agricultural products and prevented farm prices from finding a realistic level.[4]

[3] Henry C. Wallace, *The Wheat Situation: A Report to the President* (USDA, 1923), p. 32; USDA *Yearbook of Agriculture, 1923*, p. 544; and USDA *Yearbook of Agriculture, 1924*, p. 185.

[4] Murray R. Benedict, *Farm Policies of the United States, 1790–1950* (New York: Twentieth Century Fund, 1953), pp. 174–75.

When the decline failed to materialize, farmers interpreted it as a sign of permanently higher agricultural prices. Encouraged by this happy prospect, they went deeper into debt to buy still more land and machinery. Despite hints by the Agriculture Department of the possibility of imminent price adjustments, excited farmers ignored the warnings and prepared to participate in what they were confident was a promising new era for American agriculture.[5]

In the summer of 1920 agricultural prices began their inevitable decline, leaving in their wake a nation of disillusioned and bankrupt farmers. Accelerating rapidly, the downward movement had carried prices of most commodities to prewar levels and lower by the end of the year. The inability of Europe to buy American products at inflated prices was the principal cause for the precipitous drop. As a result of economic overextension during the war and unstable conditions after the armistice, the purchasing power of the Allies declined rapidly in the immediate postwar period. In early 1920, furthermore, the United States government terminated wartime credit, leaving Europe without the means to buy the food it desperately needed. Thus while the need for American agricultural products remained strong, the market for them at wartime prices disappeared. Illustrative of this fact, between the fiscal years 1919–1920 and 1921–1922 the value of farm exports dropped 50 per cent while volume actually increased slightly.[6] Postwar competition of Canada, Argentina, and Australia, coupled with the recovery of agricultural regions in Europe, also tended to force prices down.

At the same time agricultural output in the United States

[5] Henry C. Taylor, "A Farm Economist in Washington," p. 97, ms. in the Henry C. Taylor Papers, State Historical Society of Wisconsin, Madison.

[6] USDA *Yearbook of Agriculture*, 1935, p. 633. While the volume of agricultural exports declined in 1923 and 1924, it is interesting to note that it remained above prewar levels throughout the 1920's.

began to reflect the influence of the additional land which farmers had brought into production in the previous ten years. Although total land in farms increased a modest nine per cent between 1910 and 1920, the actual area devoted to crops and livestock grew considerably more. For one reason, the widespread practice of replacing horse power with tractor power freed thousands of acres of pasture. During the war high prices induced farmers to prepare unimproved or marginal land for crop production. Woodland on farms, for instance, declined by over 12 per cent in the second decade of the twentieth century, while land in cereal crops increased by almost 15 per cent, with wheat acreage expanding a disproportionate 65 per cent.[7] Through greater use of machinery, commercial fertilizer, and improved methods, moreover, producers were able to work their land more intensively and thus gain higher yields. Faced with a weakening foreign market, American farmers could ill-afford the increasing production which resulted from these changes.

Although the price decline was manifest throughout the economy, the decrease in nonfarm goods was relatively less than in agricultural commodities. Producers were thus forced to trade a larger share of their goods for the nonfarm products they needed than had been the case before the war. During the critical year of 1920 the relative value of farm commodities plummeted from 91 per cent to 65 per cent of its 1913 level. At its low point in November 1921 the purchasing power of farm commodities had fallen to 63 per cent of the prewar average.[8]

The price decline affected agriculture in yet another way. Farm indebtedness more than doubled and interest assessments grew by almost 140 per cent between 1910 and 1920. Producers,

[7] *Fourteenth Census of the United States,* Agriculture, 5, pp. 24, 700.

[8] *Agricultural Situation* (USDA), August, 1924, p. 20; George F. Warren and F. A. Pearson, *The Agricultural Situation* (New York: John Wiley and Sons, 1924), pp. 60–67; and Clarence A. Wiley, *Agriculture and the Business Cycle Since 1920* (Madison: University of Wisconsin Press, 1930), p. 15.

particularly those who had recently incurred large debts during and immediately following the war to invest in overvalued land and machinery, found themselves in a precarious situation. Having contracted obligations at inflationary levels, they were now forced to pay them off with income from depressed prices. To add to their plight, mortgaged land was in many cases now worth much less than the debt it carried due to the sharp drop in land values beginning in 1920. During the latter nineteenth century producers had experienced a similar situation as gradually falling prices worked a particular hardship on debtors. But the wide adjustments of 1920 placed mortgaged farmers in a much more severe financial predicament than that of the post-Civil War era, with the result that many simply could not meet their obligations and were ruined financially. In 1914, 5.4 per cent of the country's bankruptcy cases were brought by farmers, but by 1924 the percentage had risen to 18.7, the highest proportion in the first half of the twentieth century.[9]

Adding to their difficulties, farmers found that the easy credit available only a short time earlier had suddenly disappeared. It was virtually impossible in many parts of the country for them to obtain loans to refinance mortgages or to carry them until the anticipated rise in prices. Alarmed by the economic collapse, rural banks initiated retrenchment policies to relieve overextended positions which they had allowed to develop during the period of agricultural prosperity. They sharply reduced new loans, refused to renew expiring loans, and pressured farmers to liquidate existing loans. Reversing its wartime policy of easy credit, the Federal Reserve Board aggravated conditions by raising the rediscount rate after the price decline began.[10] As a result, farmers faced a

[9] Wallace, *The Wheat Situation,* pp. 35–36; *Agricultural Statistics* (USDA), 1936, pp. 334, 354.

[10] USDA *Yearbook of Agriculture,* 1924, p. 231. Arthur S. Link discounts the importance of the Federal Reserve Board's policy as a factor affecting agricultural conditions in the postwar period. He maintains that "Governor [William P. G.] Harding, the Board, and the Reserve

money stringency at a time when they sorely needed liberal credit.

In addition to fixed interest and mortgage payments, producers found taxes on farm land and transportation rates likewise burdensome. Property assessments on agricultural land rose an average 140 per cent between 1914 and 1923, while in Kansas the increase was 171 per cent and in several counties of Washington it exceeded 200 per cent.[11] Farmers also had to absorb substantial increases in freight rates during the same period. While railroads were under the wartime control of the government transportation charges had remained fairly constant. Upon return of the roads to private ownership in 1920, however, the Interstate Commerce Commission authorized upward adjustments in rates of from 25 to 40 per cent.[12] Coinciding with the decline in agricultural prices, higher taxes and growing freight costs represented further expenses which depressed farmers had to meet.

The toll of the postwar depression in rural areas was great. Many farmers lost their land and equipment, and those who managed to retain their property did so in many cases only through great personal sacrifice. Keeping farms intact, moreover, was no assurance of an adequate living. Due to the downward spiral, prices on agricultural commodities often failed to meet the farmers' production costs or at best offered them only minimal

banks were made the scapegoats of the agricultural depression by small local bankers who had grossly overborrowed, by demagogic politicians who were serving their own ends primarily, and by farmers in general who did not understand what had happened with regard to the credit situation, but who were suffering greatly in the economic chaos of the period." See "The Federal Reserve Policy and the Agricultural Depression of 1920–1921," *Agricultural History* 20 (July, 1946): 175.

[11] USDA *Yearbook of Agriculture*, 1924, p. 257; and Fite, *George N. Peek*, p. 12.

[12] James H. Shideler, *Farm Crisis, 1919–1923* (Berkeley and Los Angeles: University of California Press, 1957), p. 58; and James E. Boyle, *Farm Relief* (Garden City: Doubleday, Doran, 1928), pp. 16–18.

profits in return for heavy investments in both money and labor. Their increasing wealth in the first two decades of the twentieth century had enabled farmers to purchase automobiles and home appliances, install telephones and electricity, and obtain many of the luxuries of a developing industrial economy. With the depression, however, they lost many of their newly acquired conveniences and suffered declines in living standards, a situation which precipitated an erosion of rural morale and accelerated the movement of population from farming to urban areas.

Wallace understood better than many of his contemporaries the implications of the wartime boom for American farmers. Realizing that artificially high prices could not last, he admonished producers immediately following the war to cut back on output and guard against overextension. Prices were sure to fall, he cautioned early in 1919, contending that the only thing in doubt was the nature and progress of the decline. Fearful of a sudden adjustment in the economy, Wallace recommended that the government should undertake some program for gradual price adjustment. "If we don't start climbing down a rung at a time," he warned, "there will be a disturbance of some kind and someone will get pushed off the ladder to land with a thud at the bottom." Prophetically, he predicted that farmers were certain to be the first "pushed off the ladder." Anticipating that price declines would not be uniform, he also accurately foresaw that they would bear most heavily on producers. To Wallace the reasons for this were clear. Unlike other sectors of the economy, the agricultural industry lacked an understanding of the price-making process and the power to control production, both essential if farmers were to protect themselves in the midst of falling prices.[13]

Wallace correctly read the significance of the change in inter-

[13] *Wallaces' Farmer* 44 (January 17, 1919): 112 and 44 (February 7, 1919): 312.

national economic relations brought about by World War I. Until 1914 private interests in the United States had been in debt to financiers in several countries of western Europe, a condition which permitted a favorable balance of trade for the United States without disturbing international exchange, since the country's excess in export receipts could be used to service and reduce its foreign debt. But the war produced an important change in the American economic position. Emerging from the conflict as the leading creditor nation in the world, the United States found its principal customers of western Europe also its major debtors. Quick to see the significance of this development for the future of American trade, Wallace pointed out that if these countries were to be able to pay their debts, the United States would have to buy more and sell less abroad. In view of the increased dependence of farmers on the export trade, he warned, such a situation boded ill for the future of the agricultural industry.[14]

As early as December 1918 Wallace predicted that freight rate increases, certain to come once railroads were returned to private control, would work a hardship on producers. Anticipating that farmers would be ill-prepared to absorb rising transportation costs at the same time prices for their products were declining, he urged stronger control for the Interstate Commerce Commission as a check against unwarranted rate increases.[15] Also clear to Wallace were the hazards of participating in the land boom, and he complained to a friend that "our people seem to have gone wild on land." As farmers rushed to add to their holdings during and following the war, he cautioned that speculation had pushed real estate prices to unrealistic levels and advised against further purchases.[16]

[14] *Ibid.*, 43 (September 27, 1918): 1368 and 45 (January 23, 1920): 258.

[15] *Ibid.*, 43 (December 6, 1918): 1772.

[16] *Ibid.*, 44 (April 18, 1919): 763; and Wallace to Cole Ambrose, June 21, 1919, Henry C. Wallace Papers, University of Iowa Library, Iowa City, Box 2.

Once the full impact of the depression had been felt, farmers began to wonder how long it would last. Most agreed with Wallace that the agricultural crisis would be temporary and that within a year or two they could expect progress toward recovery. Assuming that the depression would follow earlier patterns, producers convinced themselves that they needed only to endure current hardships for a short time while the economy readjusted to peacetime conditions. As it turned out, they badly miscalculated and were poorly prepared to meet the prolonged period of financial distress which was to follow.

Wallace's belief that the farm crisis would be short-lived influenced his thinking on possible relief measures. While he maintained that any permanent improvement in agriculture had to come from fundamental changes within the industry itself, the editor felt that the federal government had a responsibility to help producers through the period of hardship caused by the rapid price declines. Farmers had responded to appeals for increased production during the war, he had now decided, and should not be penalized for their patriotism. Since they had no way to protect themselves from postwar fluctuations, it became the government's duty to aid them until stability had returned to the economy. Wallace suggested that this could be done through improved credit facilities, an agricultural tariff, reduction of freight rates, federal regulation of processors and marketing agencies, the lifting of restrictions on cooperatives, and stimulation of farm exports.[17]

All of these measures, to be sure, were included in the Republican agricultural plank of 1920. But Wallace was aware that inclusion in a party platform was no guarantee of enactment, and even before the election he took steps to insure that no matter

[17] *Wallaces' Farmer* 45 (December 10, 1920): 2748; and *New York Times,* March 13, 1921, p. 10. Wallace also advanced a government program of farm economics as the basis for a more fundamental modification in the agricultural industry. This aspect of his policy will be discussed in a later chapter.

which party won in November Congress would at least consider his suggestions. At his urging officers of the Iowa Farm Bureau Federation called a meeting of the state's congressional delegation to discuss farm problems. The group, with only Cummins absent from among the congressmen, met with Wallace in Detroit in the fall of 1920. As would later become clear, Cummins opposed this kind of extraparty activity and apparently chose to disassociate himself from the conference. The meeting arrived at general agreement on the need for the agricultural measures advanced in the Republican platform, and, more importantly, the congressional delegation led by Senator William S. Kenyon pledged support of the party's program and promised to work for its enactment by Congress.[18]

The Detroit meeting was followed by another after Harding's election. Assembling in Des Moines on November 13, 1920, the same group again considered the agricultural situation and possible solutions, and once more endorsed the Republican farm plank, emphasizing the need for tariff revision, improvement of rural credit, and stimulation of foreign trade. "Out of this conference," Wallace announced after the meeting, "should grow in the next Congress a group comprising the representatives from agricultural states, who will work together as a unit for laws to protect the farmer, and place first in their consideration the solution of the country's agricultural problem." [19]

Formation of such a group took place on May 9, 1921, in the Washington offices of the American Farm Bureau Federation. At the instance of Gray Silver, the federation's representative in Washington, Kenyon summoned senators from farm states to a meeting to discuss a program of agricultural relief and to agree on a strategy for bringing about its enactment. In addition to

[18] Russell Lord, *The Wallaces of Iowa* (Boston: Houghton Mifflin, 1947), p. 215.
[19] *Fort Dodge Messenger,* November 13, 1920, p. 1.

Kenyon, 11 other senators as well as A. F. Lever of the Federal Farm Board attended the session. Although Wallace was not present, he sent two representatives from his department, Henry C. Taylor and Chester Morrill.[20] The senators at the meeting formed a permanent, if unofficial, organization with Kenyon as chairman, and pledged to cooperate in support of any legislation which offered aid to distressed farmers.[21]

The "farm bloc," as the group was soon dubbed, became one of the really powerful forces in Congress in the early 1920's. Membership grew from the original 12 to a high of 22 when the bloc was at the peak of its influence. In addition, a number of senators who refused outright affiliation with the bloc consistently cooperated and voted with its members. Although made up largely of senators from the Middle West and Great Plains, the group also included a number of southern congressmen. Besides cutting across sectional lines, the bloc was bipartisan and welcomed any senator with an interest in agricultural relief. According to one of the members, there was "too much party and too little individual responsibility in Congress. An organization along the lines of the 'Agricultural Bloc' is much more logical than an organization along political lines."[22] Working closely with the Farm Bureau, the organization formulated its program in cooperation with the

[20] Memorandum, Chester Morrill to W. A. Jump, May 10, 1921, Files of the Bureau of Agricultural Economics, National Archives. The senators attending were Kenyon, George W. Norris of Nebraska, Frank R. Gooding of Idaho, Arthur Capper of Kansas, Ellison D. Smith of South Carolina, Duncan U. Fletcher of Florida, Robert La Follette of Wisconsin, Morris Sheppard of Texas, Edwin F. Ladd of North Dakota, John B. Kendrick of Wyoming, Joseph E. Ransdell of Louisiana, and J. Thomas Heflin of Alabama.

[21] *Ibid.*

[22] George W. Norris to H. E. Stockbridge, January 9, 1922, George W. Norris Papers, Manuscript Division, Library of Congress, Tray 5, Box 1. For a discussion of the farm bloc see Alice M. Christensen, "Agricultural Pressure and Government Responses in the United States, 1919–1929," Doctoral dissertation, University of California, Berkeley, 1936, pp. 63–64.

federation's Washington offices. Through a network of state and county bureaus, Gray Silver obtained information on the needs of agriculture which the bloc used to guide its work. Under the leadership of Lester J. Dickinson of Iowa, representatives in the House formed a similar group, but the membership was more fluid and it lacked the discipline of its counterpart in the Senate.[23]

[23] Gilbert C. Fite, "Activities of the Farm Bloc in the Early Twenties," Master's thesis, University of South Dakota, 1941, pp. 16–24; and John K. Barnes, "The Man Who Runs the Farm Bloc," *World's Work* 45 (November, 1922):58

Chapter V / WALLACE AND CONGRESS, 1921–1923

W HEN Wallace assumed office in March 1921 his plan of action included two separate programs. First of all, he intended to use the influence of the Agriculture Department to secure the passage of legislation which would implement several of the Republican campaign promises to farmers. Secondly, he launched a reorganization of his own division for the purpose of improving and expanding its work in agricultural economics.[1] With the formation of the farm bloc two months after the new administration took over, Wallace had a congressional vehicle through which he could work to gain the objectives of the former program.

That the farm bloc was to be a political force of some influence became obvious soon after the new Congress convened. At Harding's suggestion Massachusetts Senator Henry Cabot Lodge introduced a resolution in July 1921 for adjournment until the House had completed drafting a new tariff bill. Convinced that tariff revision should be the first order of business, the President urged Congress to give full attention to it before taking up any other legislation. No doubt he was concerned over a number of bills recently introduced by farm bloc members, upon which they were urging quick action. Although several of these measures were already on the calendar, acceptance of Lodge's resolution

[1] The agricultural economics program is covered in Chapter VI.

would have meant tacit recognition that action on the tariff bill would take precedence once the Senate resumed. Naturally opposed to such a move, the farm bloc marshalled its forces and succeeded in blocking adjournment of the Senate. After the bloc had convincingly demonstrated its strength, Harding had little choice but to concede. He sent for Senator William S. Kenyon and the two worked out a compromise in which Congress would act upon several of the more urgent farm bills and the rest would be held until later.[2]

The Kenyon-Kendrick bill, providing for close federal regulation of the meat packers and stockyards, was the first measure which the bloc chose to focus upon in the Senate. Introduced in essentially the same form as the unsuccessful Kenyon-Kendrick bill of the year before, the measure was designed to check oligopolistic conditions which had resulted from the collaboration of the country's five major meat packers.[3] Collusion among the "Big Five," as they were known, had given them extensive control over livestock marketing, transportation to stockyards, and retail outlets in addition to the meat packing industry itself. Through their domination of all functions connected with the industry, the packers had been able, among other things, to eliminate competitive bidding on livestock, which allowed them virtually to set their own prices for the stock they purchased. Under such conditions, of course, producers were at a marked disadvantage.

In an extensive investigation of the industry in 1918–1919, the Federal Trade Commission revealed the vast power of the Big Five and declared that their activities amounted to a conspiracy in restraint of trade. Yielding to the public pressure which publication of the commission's findings aroused, Attorney Gen-

[2] John K. Barnes, "The Man Who Runs the Farm Bloc," *World's Work* 45 (November, 1922): 51–52.

[3] The "Big Five" were Armour, Morris, Cudahy, Swift, and Wilson.

eral A. Mitchell Palmer brought an antitrust suit against the packers in August 1919. Early the following year, however, the Justice Department announced that the industry had consented to divest itself of its retail interests and stockyard connections, thus making it unnecessary to continue the case.[4] Bitterly opposed by farmers, the "consent decree" lent impetus to the demand for stronger regulation of the packers. When the meat industry, in seeking modification of the agreement to allow retention of certain outside interests, demonstrated its reluctance to abide by the decree, the resentment of livestock producers was hardly mitigated.[5] Little wonder, therefore, that the farm bloc gave priority to the Kenyon-Kendrick bill.

For several years before becoming Secretary of Agriculture, Wallace had urged stricter regulation of the meat industry. "Even granting the packers the purest of motives," he wrote in 1918, "it is a decidedly bad thing that such a small number of men should, without public restrictions, be able to control so closely the meat business of the country." Wallace felt that marketing facilities should be "controlled by interests which will do justice to all who use them," and he supported the recommendation of the Federal Trade Commission that the government should divest the packers of their subsidiary interests. Moreover, he warned in 1919 that "there must be much more government regulation of the packing business than there is now and the sooner this is brought about, the better it will be for all concerned." [6]

After the introduction of the Kenyon-Kendrick bill in 1920, the Corn Belt Meat Producers' Association, under Wallace's urging, passed a resolution favoring the legislation and sent repre-

[4] James H. Shideler, *Farm Crisis, 1919–1923* (Berkeley and Los Angeles: University of California Press, 1957), p. 27.

[5] *Christian Science Monitor,* September 24, 1921, p. 4.

[6] *Wallaces' Farmer* 43 (March 8, 1918): 436; 43 (April 2, 1918): 642; 43 (October 10, 1918): 1409; and 44 (July 11, 1919): 1353.

sentatives to testify before the House Committee on Agriculture and Forestry.[7] As chairman of an *ad hoc* organization called the Producers' Committee, Wallace was also instrumental in enlisting the backing of this group. Formed after the war as a data-gathering and propaganda agency to represent various livestock organizations in the Middle West, the group conferred with representatives of the meat industry several times during 1919 in an effort to secure redress of producers' grievances. Failing in this, the committee concentrated its efforts the following year on gaining adoption of the Kenyon-Kendrick bill.[8] Wallace's organizational work in behalf of the legislation and his editorial endorsement in *Wallaces' Farmer* earned him John B. Kendrick's support for the secretaryship in the hope that he would be able to rally the new administration behind strong regulation of the packers.[9]

After becoming Secretary of Agriculture, Wallace continued to be an outspoken proponent of greater federal regulation. Though he stoutly rejected suggestions that the government should take over the meat industry, he contended that it had "reached proportions where it may be considered a public utility." [10] What was more, he now opposed any move to break up the companies. Appearing before the House Committee on Agriculture and Forestry, he expressed the view that the industry had become so large and inter-related as to render dismantling unrealistic. Because the meat industry provided indispensable

[7] Wallace to John B. Kendrick, February 24, 1920; and Wallace to Gilbert N. Haugen, February 26, 1920, Henry C. Wallace Papers, University of Iowa Library, Box 2.

[8] Wallace to W. J. Carmichael, June 30, 1919; Wallace to F. A. Buechel, April 3, 1920; and "Minutes of the Producers' Committee," 1919–1920, Wallace Papers, Boxes 2, 4.

[9] For editorials on the bill see *Wallaces' Farmer* 45 (January 16, 1920): 158; 45 (February 20, 1920): 618; 45 (February 27, 1920): 692; and telegram, John B. Kendrick to A. Sykes, January 4, 1921, Wallace Papers, Box 2.

[10] *New York Times,* June 8, 1921, p. 3; and *Des Moines Register,* June 30, 1921, p. 4.

services to livestock farmers, he cautioned, "any legislation or any rules or regulations which really impair the efficiency of the packing business will cause hardship, not to the packers so much as the producers." "Real supervision" was therefore the only feasible solution to the problem of packer abuses.[11]

Wallace's testimony before the committee revealed a change in attitude since 1918 when he had endorsed the Federal Trade Commission's recommendations to divest packers of their subsidiary interests. For reasons not entirely clear, he decided that it was time the country took a "somewhat different attitude" toward business consolidation. Where such activity reduced production and distribution costs, Wallace pronounced himself in favor of it, but only "under thorough supervision and regulation." [12] Unlike most agrarian progressives, Wallace had come to reject the antitrust solution to the problems of business consolidation in favor of regulated capitalism. No champion of the simpler type of competitive economy, he adopted instead the strand of progressivism which viewed consolidation as desirable because it meant increased efficiency and the elimination of waste. In this and in other aspects of his secretaryship Wallace revealed the influence of his earlier association with and admiration for Theodore Roosevelt.

As it had done the year before, the meat industry activated its powerful lobby in opposition to the Kenyon-Kendrick bill. Although the packers succeeded in blocking passage of the measure, they failed to prevent the enactment of a weaker compromise proposal. Under the Kenyon-Kendrick bill federal licensing

[11] United States Congress, House, Committee on Agriculture and Forestry, *Hearings on H.R. 14 by Mr. Haugen, H.R. 232 by Mr. Anderson, H.R. 5034 by Mr. McLaughlin (Nebraska), and H.R. 5692 by Mr. Williams,* 67th Cong., 1st Sess., p. 233.

[12] *Ibid.,* pp. 240, 246; see also Wallace to T. W. Tomlinson, May 5, 1921; and Wallace to E. C. Roberts, May 16, 1921, Files of the Secretary of Agriculture, National Archives.

of packers was to be the principal instrument of regulation, but, much to the despair of the farm bloc, the final legislation eliminated this provision. When it became evident that their measure would not pass, however, bloc members voted for the substitute and later claimed credit for its enactment.[13]

Although Wallace much preferred the Kenyon-Kendrick proposal to the one accepted by Congress and had sought Harding's support for it, he agreed with the farm bloc that the weaker bill was better than no legislation at all. Shortly after its passage, therefore, he urged the President to approve the measure.[14] Harding's silence during the fight over the packer bills was in itself testimony to the fact that he had little interest in the regulation of the meat industry. But the bill which finally emerged from Congress was one which he could accept without serious compromise of his political philosophy and was something to which he could point as fulfillment of a campaign promise. Thus he followed Wallace's advice and signed the bill into law in August 1921 as the Packers and Stockyards Act.

The Packers and Stockyards Act vested the Department of Agriculture with considerable authority over meat and livestock marketing industries by empowering the secretary to investigate the activities of companies suspected of operating in restraint of trade. Among other things, the act required livestock shippers to charge nondiscriminatory rates and to meet minimum standards set by the government for their facilities and services. Stockyards had to operate under a competitive bidding system which insured equal treatment of all customers. If a company were found to be in violation of the law, the Agriculture secretary was to issue a cease and desist order, after which the firm would be subject to a fine for repetition of the offense. In addition, the Federal Trade

[13] *Congressional Record*, 67th Cong., 1st Sess., p. 4642; and Orville M. Kile, *The Farm Bureau Movement* (New York: Macmillan, 1921), p. 174.

[14] Wallace to Harding, August 15, 1921; and Wallace to George B. Christian, Jr., August 12, 1921, Files of the Secretary of Agriculture.

Commission retained its investigatory powers over the packers. The industry, however, might appeal the rulings of either the Secretary of Agriculture or the commission in the federal courts.

Wallace also applied his ideas on regulated capitalism to the grain trade. Since Iowa was primarily a livestock rather than cash grain state, he was less concerned with the commodity trade than the meat industry before becoming Secretary of Agriculture. But soon after taking office he recognized that wheat farmers of the Great Plains were as much at the mercy of the grain exchanges as livestock producers of the Middle West were dependent upon the packers. Still, he felt that the system of determining prices through the exchanges was the best method for the American economy, and he insisted that "we must do nothing to interfere with the orderly use of their legitimate function." To insure that the private grain trading agencies operated responsibly Wallace favored federal regulation, and when the farm bloc proposed legislation designed to check manipulation of commodity marketing he was quick to support it.[15]

Although Wallace strongly endorsed the farm bloc's Capper-Tincher bill for regulation of the grain trade, he recognized several weaknesses in the proposal and directed a member of the Agriculture Department, Chester Morrill, to study the measure and make suggestions for possible changes. Appearing before the House committee conducting hearings on the bill, Wallace recommended the modifications proposed by Morrill, most of which were accepted and eventually incorporated in the final legislation. With these revisions, the secretary felt that the bill was greatly strengthened and he actively campaigned for its enactment.[16]

[15] *New York Times,* June 8, 1921, p. 3; and *Des Moines Register,* June 30, 1921, p. 4.

[16] Morrill to Wallace, May 28, 1921; "Report on Meeting with the House Committee on Agriculture and Forestry, June 2, 1921," Files of the Bureau of Agricultural Economics, National Archives; and address by Wallace before the Millers' National Federation, Chicago, June 29,

Despite the opposition of the grain trade, the Capper-Tincher bill passed Congress in August 1921 and became law as the Grain Futures Trading Act. Less than a year later, however, the Supreme Court invalidated the law. To discourage speculation the act had placed a prohibitive tax on grain sold for future delivery by anyone other than the producer, which the court judged was an unwarranted use of congressional taxing power. With the help of Wallace and the Agriculture Department, the Capper-Tincher bill was redrafted without the taxing provision and passed in September 1922.[17] The Grain Futures Act, as the new law was called, gave the Secretary of Agriculture power to regulate commodity exchanges for the purpose of preventing market manipulations and monopolistic practices. Well satisfied with the revised law, Wallace felt it could effectively accomplish its purpose without interfering with normal grain transactions.[18]

On one important bill the secretary felt compelled to part with the farm bloc. In view of the growing domestic surplus in agricultural products, farm representatives reasoned that if only the portion of production not needed at home could be channeled abroad, prices on farm goods would return to a more realistic level in relation to other commodities. Of several measures introduced into Congress in the first two years of Harding's administration to increase farm exports, a bill authored by George W. Norris of Nebraska attracted the strongest support as well as the backing of the farm bloc. This measure proposed the creation of a government corporation, financed through federal funds, which would be empowered to purchase agricultural com-

1921, copy in Edwin T. Meredith Papers, University of Iowa Library, Iowa City, Box 10.

[17] United States Congress, House, Committee on Agriculture and Forestry, *Hearings on an Act for the Prevention and Removal of Obstructions and Burdens upon Interstate Commerce in Grain, by Regulating Transactions on Grain Future Exchanges, and for other Purposes,* 67th Cong., 2nd Sess., p. 47.

[18] *Des Moines Register,* June 1, 1922, p. 1 and November 1, 1922, p. 8.

modities on the domestic market and dump them on the foreign market. Also included in the bill were a provision for the extension of credit on easy terms to foreign buyers and a plan for the corporation to use the government-owned merchant marine, which had been built during the war, to transport goods at low rates. Another cost-cutting feature would have made it possible for the agency to deal directly with foreign governments, corporations, and individuals without working through middlemen in the United States.[19]

If Harding remained silent to the Kenyon-Kendrick bill, he showed no inclination to do the same on Senator Norris' proposal. While he could consent to federal regulation of food processors and handlers, a government owned corporation engaged in the export business was another matter entirely. His concept of the proper sphere of federal power might be expanded, but not far enough to sanction that much government involvement in the economic process. Fearful that the farm bloc might succeed in pushing the Norris bill through Congress, Harding had Secretary of Commerce Herbert C. Hoover and Eugene Meyer, director of the War Finance Corporation, draw up an alternative bill which the administration advanced as a substitute.[20] The measure provided for enlargement of the working capital of the recently revived War Finance Corporation for the purpose of financing farm exports and alleviating the money stringency in rural areas. The corporation was to be empowered to make loans not only to exporters but to banks and agricultural cooperatives as well, although it would not be permitted to deal directly with foreign agents.

In full agreement with the President, Wallace insisted that

[19] See George W. Norris, *Fighting Liberal* (New York: Macmillan, 1945), pp. 279–80.

[20] Hoover to Harding, August 8, 1921, Warren G. Harding Papers, State Historical Society of Ohio, Columbus, Box 5; and Hoover to Harding, August 18, 1921, Herbert C. Hoover Papers, Herbert C. Hoover Presidential Library, West Branch, Iowa, 1-I/4.

government participation in the export business held the same threat to American political and economic institutions as federal ownership of the railroads, the meat industry, or grain exchanges. All such solutions to farm problems he rejected as "visionary schemes" which in the long run would bring far greater damage than relief.[21] Confirming Wallace's feelings on the measure, an Agriculture Department committee studied the Norris bill and deemed it "a very dangerous experiment for the government to undertake and one which would not appeal to the sober business sense of even the more intelligent class of farmers." The committee also declared that from purely practical considerations there was no need for the corporation because the financing and handling of agricultural exports were already well provided for by the War Finance Corporation and private agencies.[22] Another member of the department reminded Wallace, moreover,

[21] "Report of the Secretary," USDA *Yearbook of Agriculture, 1921*, p. 15. Donald R. Murphy remembers that in a talk with Wallace in December 1922 the secretary indicated that he favored the Norris bill. This would contradict Wallace's indirect reference to the proposal in his 1921 report. At the hearings on the Norris bill after it was reintroduced in 1924, furthermore, Wallace was reported to have opposed the measure in 1921. Benjamin C. Marsh, president of the Farmers' National Council, also remembered in 1924 that Wallace had spoken against the Norris bill three years earlier. The Wallace family biographer and two agricultural historians of the period also maintain that the secretary opposed the measure as too radical. The problem is complicated by the fact that Wallace made no public statement directly on the Norris bill, as he did on several other measures advanced by the farm bloc. Interview with Donald R. Murphy, Des Moines, Iowa, August 29, 1965; Murphy to the author, January 11, 1966; United States Congress, Senate, Committee on Agriculture and Forestry, *Hearings on Purchase and Sale of Farm Products,* 68th Cong., 1st Sess., p. 324; Farmers' National Council press release, undated (early 1924), copy in files of the Secretary of Agriculture; Russell Lord, *The Wallaces of Iowa* (Boston: Houghton Mifflin, 1947), p. 222; Shideler, *Farm Crisis,* p. 160; and Gilbert C. Fite, *George N. Peek and the Fight for Farm Parity* (Norman: University of Oklahoma Press, 1954), p. 88.

[22] Memorandum, E. D. Ball to Wallace, June 21, 1921, Files of the Secretary of Agriculture.

that low farm prices were not due to a decline in export volume, which had remained high since the war, and he echoed the committee's report by cautioning that the "entry of the Government into business should be undertaken only in very serious emergency."[23] Despite the attitude of the department and his own aversion to the Norris bill, Wallace was apparently reluctant to alienate the farm bloc by publicly announcing his opposition, for he refrained from any direct statements on the measure and only alluded to it in his annual report of 1921.

Minnesota Senator Frank B. Kellogg introduced the administration's proposal into Congress. Drafted in secret without consultation with any of the members of the Senate Agriculture Committee, the Kellogg bill came as a surprise to the farm bloc. To make matters worse, backers of the measure used a procedural trick to gain recognition of the Minnesota senator so that he could present the measure.[24] Soon after the bill was revealed, Harding dispatched a note to the Senate urging quick passage and stepped up his campaign against the Norris proposal.[25] Rapid action by the administration undermined the movement for Norris' bill and resulted in the enactment of the Kellogg substitute. When it became apparent that their measure would not pass, several members of the farm bloc voted for the Kellogg bill with the idea that improved rural credit would provide at least some assistance to depressed farmers. Confident that with the solid and continued support of the bloc his bill would ultimately have been enacted, Senator Norris was incensed at this desertion.[26] Harding's strategy in advancing the substitute proposal,

[23] Memorandum, George Livingston to Wallace, June 16, 1921, *ibid.*

[24] *New York Times,* July 27, 1921, p. 8; and William Allen White, *A Puritan in Babylon* (New York: Macmillan, 1938), p. 237.

[25] *Congressional Record,* 67th Cong., 1st Sess., p. 4299.

[26] Norris to J. T. James, January 12, 1922; Norris to Harry N. Owens, January 14, 1922, George W. Norris Papers, Manuscript Division, Library of Congress, Tray 5, Box 1; and Norris, *Fighting Liberal,* pp. 281–83.

therefore, not only succeeded in defeating the Norris bill but laid the seeds of dissension which would eventually contribute to the decline of the farm bloc.

Although Wallace had nothing to do with the administration's action on the measure, he was well pleased with the passage of the Agricultural Credits Act, as the Kellogg bill was called. For one thing, he attributed the slight improvement in agricultural conditions in early 1922 in part to the enlargement of the power of the War Finance Corporation. In his annual report of the same year, he maintained that the activities of the corporation "undoubtedly saved many thousands of farmers from bankruptcy and hundreds of banks in agricultural states from passing into the hands of receivers." Not only had increased federal credit enabled farmers to refund debts at reduced interest rates, he explained, but government action forced private agencies to lend more readily and at lower rates. The Agricultural Credits Act might not have solved the farm crisis, Wallace later reflected, but it helped to temper its severity.[27]

The farm bloc was instrumental in bringing about the enactment of yet another piece of agricultural legislation. Passed in February 1922, the Capper-Volstead Act exempted farm cooperatives from prosecution under the antitrust laws. The formation of nationwide marketing associations had been retarded because of the fear among certain groups that antitrust restrictions prohibited this kind of cooperative activity. With farmers anxious to have their organizations freed of legal hindrance, the farm bloc responded by giving high priority to the Capper-Volstead bill.

As early as 1919 Wallace had urged the enactment of an amendment to the antitrust laws exempting agricultural associa-

[27] *Washington Post,* March 16, 1922, p. 5; "Report of the Secretary," USDA *Yearbook of Agriculture,* 1921, p. 12; and Henry C. Wallace, *Our Debt and Duty to the Farmer* (New York: Century, 1925), pp. 105–7.

tions. Noting that labor unions were not subject to antitrust restrictions, the editor insisted that farmers should be provided the same consideration. A year later bills were introduced into both houses of Congress to accomplish this purpose but no action was taken on them. In support of these measures Wallace told readers of his journal that "the legislation most needed is that which will enable the farmer to form marketing organizations without running the risks of federal prosecution." [28] As Secretary of Agriculture, he continued to recommend such legislation and firmly endorsed the Capper-Volstead bill when it appeared in Congress. In many ways his views on agricultural cooperatives were similar to those on business consolidation. Where antitrust laws prevented the formation of organizations which promised greater efficiency and the elimination of waste, he believed that they should be waived. The advantages of a freely competitive society, according to Wallace, were not so great that the benefits of cooperation should be sacrificed for them. [29]

Wallace had to use little persuasion in securing Harding's acceptance of the Capper-Volstead bill. Outside of tariff revision, encouragement of cooperation was the solution to agricultural problems most frequently mentioned by the President. Before taking office he announced that "an important part of my plan to help farmers is Cooperation," and in his second address to Congress on December 6, 1921, Harding proposed that "every proper remedy should be given to the cooperative marketing programs." [30] Representing a program of self-help which required little or no aid from the federal government, the formation of farm associations understandably appealed to Harding. If he could encourage such a movement merely by stretching his

[28] *Wallaces' Farmer* 44 (May 16, 1919): 1065 and 46 (February 11, 1921): 278.

[29] Wallace, *Our Debt and Duty to the Farmer,* pp. 157–59.

[30] *Farm Journal,* December, 1920, p. 7; and *Congressional Record,* 67th Cong., 2nd Sess., p. 54.

definition of "normalcy" to include antitrust exemptions for co-operatives, this was small sacrifice indeed, and he signed the Capper-Volstead Act into law with no hesitancy.

Secretary Wallace was also interested in the efforts of the farm bloc to place an agricultural representative on the Federal Reserve Board. Though not as critical of the board as many farmers, he saw its failure to develop a well-established, constructive policy as contributory to the price instability of the postwar period. He charged that the severe credit stringency in rural areas could be attributed in part to the absence of a competent farm representative on that body.[31] It was vital to rural interests, Wallace thought, to have at least one member of the board "who has an understanding and an appreciation of agriculture, and who can therefore speak with some authority as to the possible effect upon agriculture, and especially upon the general level of agricultural prices, of any policy . . . [adopted] by the Federal Reserve System." [32] Agreeing with Wallace's assessment, the farm bloc proposed an amendment to the Federal Reserve Act specifically providing for an agricultural representative on the board. Although Harding objected to this invasion upon his appointive powers, pressure from farm representatives and the attitude of his Agriculture secretary convinced him that he should not reject the measure outright. Thus he consented to an increase in the number of board members from five to six, but refused to agree to a stipulation regarding the qualifications of the additional member. Unofficially, however, he promised the bloc that he would place a farmer in the new position.[33] In June 1922 the amendment passed Congress, and Harding appointed Milo D. Campbell, a dairy cooperative leader from Michigan, to the board. But Campbell died a few months later and his place was

[31] *Wallaces' Farmer* 45 (October 8, 1920): 2354 and 45 (October 22, 1920): 2469.

[32] Wallace to John H. Finley, December 26, 1922, Files of the Secretary of Agriculture.

[33] *Washington Post,* January 19, 1922, p. 6.

taken by a nonfarm appointee. Wallace and the farm bloc, therefore, gained little for their efforts.[34]

In 1922 the power of the farm bloc declined steadily and by the end of the year the organization enjoyed little of its former influence. An important factor in its decline was the resignation of Kenyon from the Senate. Seeking a way to undermine the strength of the bloc, Harding conceived the idea of depriving the group of its leadership. In October 1921 he offered Kenyon the federal judgeship of northern Iowa, but the senator turned down the post. Convinced that Kenyon's resignation would be unfortunate for agricultural interests, Wallace quietly urged him not to leave Congress and was much relieved when he rejected the appointment.[35] Early the following year, however, the President tried once again by offering the farm bloc leader an even more attractive position as judge of the Circuit Court of Appeals of the Eighth District. Much to the disappointment of Wallace and congressional farm representatives, Kenyon decided to accept the second offer. When he left in February 1922 to take his new position, leadership of the farm bloc fell to Arthur Capper of Kansas. While Capper was an able organizer, he lacked Kenyon's ability as a floor leader, and without the aggressive leadership of the Iowa senator the bloc lost much of its dynamism in the Senate.

Another factor in the decline of the farm bloc was internal disagreement. Besides the lack of accord on the Norris bill there were differences over tariff and credit policy. Comprised of both high-tariff western Republicans and low-tariff southern Democrats, the farm bloc could not agree on the question of protectionism, and members voted individually or as part of other groups on tariff legislation.[36] On the matter of rural credits, to

[34] Shideler, *Farm Crisis,* p. 172.
[35] *New York Times,* October 5, 1921, p. 1.
[36] Arthur Capper, *The Agricultural Bloc* (New York: Harcourt, Brace, 1922), p. 117. A group of 24 Republican senators formed a tariff bloc

be discussed later, there was considerable dissension over the kind of assistance needed and over which agency should handle it. By the time the Intermediate Credits Act was passed in March 1923, this question had so divided the bloc that its usefulness was at an end.[37]

Perhaps the main reason for the failure of the farm bloc to maintain its position, however, was the formation of a new bloc. Congressional elections of 1922 returned a large group of insurgent Republicans, men dissatisfied with the piecemeal approach of the farm bloc and anxious to make a more aggressive and broader attack on the postwar depression. Shortly after the elections Senator Robert M. La Follette of Wisconsin called a conference to convene in Washington on December 1, the purpose of which was to discuss plans for the cooperation of "progressives" in Congress. Representing mainly agriculture and labor, the meeting decided to form a new bipartisan group to work for the enactment of "progressive legislation." Calling themselves the "progressive bloc," members pledged support of a program which included not only agricultural legislation but a wide range of relief and reform measures.[38] The new organization had broader appeal than the single-interest farm bloc and was thus able to develop greater support in Congress. By the end of 1922, furthermore, the original bloc had succeeded in passing most of its program and had little more to offer, while the pro-

under the leadership of Frank R. Gooding of Idaho to work for higher agricultural duties. It included many of the western members of the farm bloc. The organization lacked the cohesion of the main group, probably because, as will be pointed out, there was really no need for a tariff bloc in the Sixty-seventh Congress.

[37] Peter Norbeck to P. W. Dougherty, December 16, 1922; and Norbeck to S. W. Clark, January 29, 1924, Peter Norbeck Letters, Western Historical Manuscripts Collection, University of Missouri Library, Columbia, Box 1.

[38] *New York Times,* November 19, 1922, p. 3; *Fort Dodge Messenger,* December 1, 1922, p. 1; and Belle C. and Fola La Follette, *Robert M. La Follette* (New York: Macmillan, 1953), pp. 1066–67.

gressive bloc was promising a more comprehensive and militant approach to agricultural problems. As the new group superseded the farm bloc in Congress, the more radical members of the original organization—men like La Follette, Norris, and Edwin F. Ladd—went over to the progressive bloc, while the rest returned to the ranks of their own parties.

Wallace regretted the decline of the farm bloc. For one reason, he felt that under its leadership "Congress . . . made a very unusual record in trying to be intelligently helpful in meeting the severe agricultural depression which . . . is now playing a large part in our industrial depression. . . ." Its program he praised as an "intelligent recognition of the inconvertible fact that national welfare depends upon a sound agriculture." Also important to him was the fact that the accomplishments of the bloc went a long way toward fulfilling the promises of the Republican platform and carrying out his own ideas on agricultural relief. Had the organization become a permanent congressional organization, as some expected it would, Wallace would indeed have been well pleased.[39]

Wallace's support of the farm bloc placed him in a delicate position within the administration. Unreservedly opposed to the group, Harding maintained that it undermined party discipline and interfered with the carrying out of Republican campaign promises. Nebulous as they were, the President had his own ideas on how the farm situation should be handled, and he resented being pressured by a group made up and led largely by members of his own party. Beyond that, bloc activity of any kind violated his sense of political order. Having risen in Republican ranks as a party regular, Harding could not fathom the thinking of those

[39] "Report of the Secretary," USDA *Yearbook of Agriculture,* 1921, p. 15; address before the Boston Chamber of Commerce, December 19, 1921, copy in Henry C. Taylor Papers, State Historical Society of Wisconsin, Madison; and Wallace to Harding, January 10, 1922, Harding Papers, Box 197.

who chose to work outside of organizational channels. "There is vastly greater security, immensely more of the national viewpoint, much larger and prompter accomplishment," he pompously lectured Congress, "where our divisions are along party lines, in the broader and loftier sense, than to divide geographically, or according to pursuits, or personal following." [40] Still, Wallace succeeded in gaining Harding's endorsement, or at least tacit approval, of much of the farm bloc's program.

Wallace held quite a different attitude toward the progressive bloc, a group he viewed as dangerously radical and irresponsible. As editor of *Wallaces' Farmer,* he had advised producers against entering into political alliances with labor groups. "Farmers should remember," he cautioned in 1919, "that the extreme, radical element which stands for socialism, and worse, is growing steadily stronger in the labor organizations." He further warned that "when a farm organization is tied up with a labor organization, the farmers who belong to that organization are betrayed." [41] In 1922 he was dismayed to find the beginnings of such a political coalition in the progressive bloc, attracting as it was the support of both farm and labor groups. If Wallace had any doubts about the nature of the new bloc, they were soon dispelled by Smith W. Brookhart's affiliation with the organization. Cummins' 1920 primary opponent had run successfully for the Senate seat vacated with the resignation of Kenyon and promptly joined forces with Republican insurgents upon his arrival in Washington.[42] Wallace's response to La Follette's call for a progressive conference was to summon several Republican senators led by Charles L. McNary of Oregon and James E. Watson of In-

[40] *Congressional Record,* 67th Cong., 2nd Sess., p. 54.

[41] *Wallaces' Farmer* 44 (October 31, 1919): 2170 and 44 (December 5, 1919): 2406.

[42] Wallace supported Brookhart in 1922, but only because he was the Republican candidate. *Des Moines Register,* June 13, 1922, p. 1; and Murphy Interview.

diana to a farm policy meeting on the same day as the conference. After the session the group pledged its support of the administration's agricultural program and announced its opposition to any radical approach to agrarian problems.[43]

In the opinion of Secretary Wallace, agriculture had gained a good deal in the first two years of Harding's administration, but the rise of the progressive bloc was an ominous sign. Not only was the new group threatening to undermine the work of the farm bloc; worse yet, it represented a sharp leftward turn among the organized forces of agriculture in Congress.[44]

If Harding made his position clear on any of the proposed solutions to the agricultural problem, it was on the question of the tariff. Increased protection for farm products was the one remedy which he accepted without qualification. Although the President hesitated to adopt, and even opposed in part, the program of the farm bloc, he led the way in bringing about tariff revision on agricultural commodities. In his first address to Congress on April 12, 1921, only one definite proposal for alleviating the farm problem emerged from a welter of meaningless phrases on the postwar depression. "It would be better," the President announced, "to err in protecting our basic food industry than paralyze our farm activities in the world struggle for restored exchanges." [45]

Early in his administration Harding expressed interest in an emergency tariff on farm products similar to the one vetoed by Woodrow Wilson shortly before he left office. Harding's predecessor had rejected the measure because of his belief that it would be impossible for Europe to pay its debts and recover eco-

[43] *New York Times,* December 1, 1922, p. 1; and Peter Norbeck to C. M. Henry, December 2, 1922, Norbeck Letters, Box 1.

[44] Wallace's fears were unfounded. The progressive bloc gave only passing attention to agricultural problems, and other than to keep the Norris bill alive concerned itself little with farm legislation.

[45] *Congressional Record,* 67th Cong., 1st Sess., p. 170.

nomically if the United States erected additional trade barriers. Less sensitive to the problems of international payments, Harding applied his traditional Republican reasoning to the farm question and decided that Wilson had vetoed exactly the legislation most likely to aid agriculture. Only two weeks after he entered office the President asked for congressional action on a new emergency tariff bill.[46]

Although not as aggressive a proponent of protectionism as Harding, Wallace came to accept increased tariffs on farm products as a partial solution to the agricultural crisis. Granting that before the war farmers had derived small benefit from protective tariffs while business interests flourished behind trade barriers, he contended that the rising competition of Canada, Australia, Argentina, and other agricultural countries had created a situation in which protection could be of real benefit to American farmers. Besides, Wallace informed his readers in 1920, "if the manufacturers are going to be protected the farmers must be protected." [47] In an Agriculture Department press release soon after taking office, the new secretary flatly stated that "practically all agricultural products should at once be given adequate protection against foreign competition," and in his first major address as a member of the Harding administration he explained that "[farm] prices are depressed not alone because of inactive business conditions at home, but because of the importations of competing foreign products." As a factor contributing to the need for higher agricultural duties, Wallace cited the recent increase in freight rates which, in effect, created "a differential against our own farmers in favor of the farmers of foreign nations with whom they must compete." [48] The remedy, as he saw it in 1921, was immediate enactment of Harding's emergency

[46] *New York Times,* March 19, 1921, p. 1.

[47] *Wallaces' Farmer* 45 (October 22, 1920): 2469.

[48] USDA press release, March 31, 1921, copy in the files of the Secretary of Agriculture; and address before the New York Academy of Political Science, New York, April 28, 1922, copy in the Taylor Papers.

tariff and a substantial increase in farm duties in the permanent tariff on which the House was beginning work.[49]

The heavily Republican Sixty-seventh Congress which convened on April 11, 1921, needed little encouragement from Harding and Wallace on the question of protection. Early in the Wilson administration Democrats, with support from insurgent Republicans, had passed the Underwood-Simmons Tariff, which represented the first significant lowering of rates since before the Civil War. Anxious to correct what they felt were the mistakes of the previous administration and to return the country once again to a high tariff policy, Republicans in Congress acted quickly on Harding's request for an emergency tariff on agricultural products. Traditionally high-tariff eastern Republicans joined with recently converted western Republicans to enact the measure into law before the end of May.[50] Outside of the South, which maintained its historic position in opposition to protectionism, farmers generally supported the bill. The cooperation of representatives from industrial areas in support of the Emergency Tariff virtually committed farm representatives to the permanent tariff, which the House was then writing and which would prove to be favorable to manufacturing interests. It ruled out, in short, the kind of agrarian opposition that Republicans encountered in 1909 when they passed the Payne-Aldrich Tariff. "It remains to be seen," former Agriculture Secretary Edwin T. Meredith ominously wrote shortly after passage of the emergency bill, "whether other interests were as anxious to see the farmers get tariff protection equal to that afforded manufacturers as they were to get the farmers in a place where they could be used to help increase tariffs already in existence." [51]

Providing for prohibitive rates on a wide range of farm prod-

---

[49] *Des Moines Register*, April 1, 1921, p. 1.

[50] *Congressional Record*, 67th Cong., 1st Sess., p. 3084.

[51] F. W. Taussig, "The Tariff Act of 1922," *Quarterly Journal of Economics* 36 (November, 1922): 5; and *Successful Farming*, July, 1920, p. 6.

ucts, including wheat, corn, meat, wool, sugar, and cotton, the Emergency Tariff in effect placed an embargo on many agricultural imports. Anticipating that the permanent tariff would shortly be completed and passed, Congress scheduled the new duties to last for only six months. When it was seen that the pending measure would not be ready as soon as expected, the Emergency Tariff was extended in November until enactment of the permanent bill.

When the second session of the Sixty-seventh Congress convened in December 1921, the new tariff bill was still in the Senate Finance Committee. In his address to Congress on December 6, 1921, Harding urged quick action on the measure, lecturing congressmen that prompt enactment was essential for the stabilization of domestic industry and agriculture as well as for the formulation of trade relations with other countries.[52] Like most tariff bills, however, this one stubbornly resisted quick action. Although the House had completed and passed a permanent tariff bill by the end of July 1921, the Senate Finance Committee debated the matter until the following April, when it finally reported the measure with over 2000 amendments. In revising the House version of the bill, the committee made good use of the services of the Department of Agriculture. At the request of the Senate, Wallace supplied rate experts and extensive data to help in the formulation of the duties on farm products, with the result that the agricultural section of the Senate's version was more carefully worked out than that of the House bill, which had merely incorporated the duties provided for in the Emergency Tariff.[53]

Wallace supported the Senate amendments affecting farm

[52] *Congressional Record,* 67th Cong., 2nd Sess., p. 53.
[53] Henry C. Taylor to Wallace, October 8, 1921, Taylor Papers; and "Report of the Secretary," USDA *Yearbook of Agriculture,* 1924, p. 24.

products because they provided for higher rates than the House bill on several severely depressed commodities. Of the opinion that the Emergency Tariff had failed to give adequate protection to the producers of certain products, the secretary was confident that the permanent bill as amended would correct these deficiencies. Whatever doubts he might have held previously on the value of agricultural tariffs, by 1922 Wallace had clearly adopted the protectionist rationale and applied it to the farm crisis.

The two houses of Congress finally agreed on a uniform bill, which they passed in late summer 1922. Approved by Harding on September 19, the Fordney-McCumber Tariff, as the new law was called, was an apparent victory for agricultural interests. Placing farm duties as high and in many cases higher than those of the Emergency Tariff, the new act taxed every conceivable agricultural product, including reindeer meat and acorns. In addition, farm implements, binder twine, potash, and several other articles necessary to producers remained on the free list. But for every gain agriculture made, other interests received more, and in the last analysis advantages to business far outweighed those to farmers. "His [the farmer's] request for protection," observed the *Country Gentleman,* "has merely served as an excuse for other interests to get more protection than they deserve." [54] The Farm Bureau calculated that under the new tariff farmers gained $125,000,000 per year in income, but had to pay an additional $426,000,000 for the goods they purchased.[55]

Of still greater significance, the Fordney-McCumber Tariff failed to remedy depressed farm prices, except perhaps on wool and sugar, and even on these products the benefit derived from protection was questionable. By September 1922 agricultural commodities had reached their low point, and the increased du-

---

[54] *Country Gentleman,* October 21, 1922, p. 10.
[55] *American Farm Bureau Federation News Letter,* January 11, 1923, p. 2.

ties of both the Emergency and Fordney-McCumber Tariffs appeared to have little effect on prices.[56] The reason for this is not difficult to discover. Most farm goods were produced in surplus of domestic need and no amount of protection could benefit products a large part of which had to be sold on the highly competitive international market. The Fordney-McCumber Tariff, furthermore, adversely affected the export of American farm goods by forcing European countries to seek sources of food and fiber which offered more favorable terms of exchange. Between the fiscal years 1922–1923 and 1924–1925 the volume of agricultural exports, which had remained strong since the war, declined by almost 25 per cent.[57] While most farm leaders, including Wallace, failed in 1922 to foresee the effect of a high-tariff policy on American agriculture, they were soon to recognize the inconsistency of advocating protection for surplus products and to turn to a more complicated, if no more effective, plan to fight the postwar crisis.

Wallace's son, Henry Agard, was one of the few in 1922 who read clearly the significance of protectionism for American agriculture. He maintained that a return to a high-tariff policy would force European countries either to satisfy their own needs for farm products or to turn elsewhere, because the means to pay for agricultural imports from the United States would be denied them due to American tariff barriers. With little opportunity to export farm commodities, it would become necessary to redirect the country's agricultural industry along lines of self-sufficiency, a task which only the federal government could successfully carry out. To accomplish this would require the elimination of agricultural surpluses, an end to the dependence on foreign markets, and the channeling of production into areas of greatest domestic need—a formidable task indeed. There was another

[56] *Agricultural Situation* (USDA), October, 1924, p. 22.
[57] USDA *Yearbook of Agriculture,* 1935, p. 633.

possibility, however, and one which conformed much better to Henry Agard's agrarian sympathies. Why not place low tariffs or none at all on the manufactured goods which Europe would have to sell to the United States in order to pay its war debt and purchase American farm goods? "Certain American industries are bound to be crushed if we follow this policy," he admitted to his father, "but the net results to the consumer . . . will be good." It was imperative, he felt, that the United States choose one of these alternatives—either agricultural self-sufficiency or lower tariffs—for it could not long continue to follow a policy of high tariffs in the midst of farm surpluses without serious economic consequences.[58]

The elder Wallace could not shed his Republican mantle so easily. Despite an ingrained resentment of big business, he was not as willing as his son to sacrifice the nation's industry to the needs of agriculture. Nor did he agree that the government would have to take on responsibility for redirecting agricultural production, confident instead that farmers, protected by high tariffs, could adjust their own output to domestic demand and thus avoid the economic consequences of which Henry Agard had warned. After all, it was not as if the agricultural depression were going to be permanent. All farmers needed from the tariff was stopgap relief to enable them to survive a temporary downturn in the economy.

Secretary Wallace and his department had more to do with the formulation and enactment of the Intermediate Credits Act of 1923 than with any other piece of farm legislation in the early 1920's. During the 1920 campaign he had cited the need for improved rural credit facilities and, as noted, was instrumental in committing the Republican party to a promise to seek remedial legislation. Conceding that the Farm Loan Act of 1916 provided

[58] H. A. Wallace to H. C. Wallace, January 3, 1922; see also H. A. Wallace to Taylor, November 3, 1926, Taylor Papers.

an important source of credit, he maintained that it was only a beginning in the solution of rural money problems. In providing long-term loans at reasonable interest rates, the act was of particular help to new farmers just entering the business and to those interested in expanding their operations. The Federal Reserve Banks, on the other hand, were empowered to rediscount agricultural paper of member banks for periods of up to six months and thus supplied an equally important source of short-term credit. What was most needed in the postwar period, to Wallace's thinking, was a "workable plan for securing credit which will enable the farmer . . . to hold his crops and market them more evenly thru the year." [59] The six-months loans of the Federal Reserve System were not long enough and it was impractical for producers to incur long-term commitments through farm loan banks for the purpose of withholding commodities.[60] Wallace complained, moreover, that the country's private credit system had been devised to meet the requirements of business and commerce, both of which operated on shorter cycles of turnover than agriculture, with the result that "the forms of short-time credit upon which he [the farmer] is obliged to rely often force him to sell his crops and live stock at a severe sacrifice." Thus he urged Congress to establish a permanent agency to provide intermediate loans of from six months to three years for agriculture.[61]

As early as September 1921 the Department of Agriculture had prepared a tentative credit bill, but Congress was too much

[59] *New York Times*, July 27, 1920, p. 3; and *Wallaces' Farmer* 45 (October 29, 1920): 2517.

[60] Federal Reserve Banks were generally opposed to making extensions on agricultural loans, and private agencies were reluctant to extend on their own. Chester Morrill to Henry C. Taylor, September 3, 1921; George E. Rommel to Wallace, January 18, 1922, BAE Files; and L. H. Goddard to C. W. Pugsley, December 2, 1922, Files of the Secretary of Agriculture.

[61] "Report of the Secretary," USDA *Yearbook of Agriculture*, 1922, p. 15.

involved with other farm legislation and the permanent tariff to consider the measure. In June of the following year Wallace sent a copy of a revision of the same bill to Representative Sydney Anderson of Minnesota, chairman of the Joint Commission of Agricultural Inquiry which had been created by Congress in June 1921 to study the farm depression.[62] In its report of October 15, 1921, the committee had recommended creation of a federal intermediate credits system for agriculture. Since the department's revised bill followed closely the suggestions of the report, Wallace apparently felt that the chairman would be the logical one to direct it in Congress.[63] Anderson, in collaboration with Representative Irvine L. Lenroot of Wisconsin, subsequently introduced a credit bill in July 1922 based upon the recommendations of his committee and those of the Agriculture Department. Wallace had hoped for action before the summer adjournment and was disappointed when the measure was laid aside; but he was pleased with Anderson's promise to push the legislation as soon as Congress reconvened in December.[64]

Wallace worked long and hard in soliciting support for the bill. Besides Anderson, he sent copies of the department's proposal to several other congressmen and requested their consideration of the measure. He met with numerous agricultural representatives, both in and out of Congress, to discuss the farmers' need for intermediate-term loans.[65] Since the Department's legislation provided for extension of the duties of the Farm Loan Board to

[62] H. C. Taylor to Wallace, September 7, 1921, BAE Files; and Wallace to Anderson, June 2, 1922, Files of the Secretary of Agriculture.

[63] United States Congress, *Report of the Joint Commission of Agricultural Inquiry,* House Report 408, 67th Cong., 1st Sess., pt. 2, pp. 8–11.

[64] Wallace to Charles E. Collins, September 11, 1922; Wallace to Anderson, August 19, 1922; and Anderson to Wallace, September 11, 1922, Files of the Secretary of Agriculture.

[65] Wallace to Charles E. Collins, September 11, 1922; Wallace to Frank R. Gooding, November 29, 1922; Wallace to James E. Watson, December 2, 1922; and Wallace to Harding, December 3, 1922, *ibid.*

cover the proposed intermediate credit system, Wallace also sought the board's support. At first opposed to the broadening of their responsibilities, the board members eventually accepted Wallace's argument regarding both the need for further rural credit facilities and the logic in placing the management under their control.[66]

A slightly revised version of the Lenroot-Anderson bill was introduced soon after the Sixty-seventh Congress reopened for its final session.[67] The measure provided for the creation of an intermediate credit system with agencies in 12 districts corresponding to those of the federal farm loan banks. Although the new system was to be administered by the Farm Loan Board, it was to have segregated assets and liabilities and to remain separate from the land banks. Under the Lenroot-Anderson bill, the intermediate credit banks were to be authorized to make loans to farm cooperatives and to rediscount the agricultural paper of banks and other private institutions for periods of six months to three years.

At the same time Senator Arthur Capper of Kansas and Representative Louis T. McFadden of Pennsylvania introduced an alternative intermediate credits measure. Drawn up by Eugene Meyer of the War Finance Corporation, the Capper-McFadden bill was based on the conviction that the Federal Reserve System was capable of meeting the needs of farmers if its powers were modestly broadened.[68] The measure recommended that federal reserve banks be permitted to extend the discount period on agri-

[66] Wallace to C. E. Lobdell, August 8, 1922; and Wallace to Harding, December 3, 1922, *ibid.*

[67] Some modification was made in the original bill at the suggestion of the Agriculture Department. Wallace to Frank R. Gooding, November 29, 1922, *ibid.*

[68] Wallace to Frank McPherrin, January 4, 1922, *ibid;* and *Washington Post,* December 31, 1922, p. 4.

cultural paper from six to nine months and that capital requirements for membership of rural banks in the system be lowered. There was also a provision for the incorporation of private credit agencies under federal charter specifically to service the needs of farmers.

Considerable division of opinion developed over the merits of the two credit bills. In the cabinet Wallace and Secretary of Commerce Hoover found themselves in rare agreement in support of the Lenroot-Anderson bill, while Secretary of the Treasury Andrew W. Mellon lent his prestige as an astute businessman to the Capper-McFadden measure.[69] Badly split on the question, the declining farm bloc found a majority of its members in favor of the Lenroot-Anderson bill, despite Capper's efforts to rally the organization behind his legislation.[70] With eastern fiscal conservatives favoring control of intermediate credits by the dependable Federal Reserve Board and rural representatives of the opinion that the Farm Loan Board would be more sympathetic to the needs of farmers, Congress in general was similarly divided on the two measures. The Lenroot-Anderson bill enjoyed wide support among agricultural interests outside of Congress, although the big livestock producers of the Far West were a significant exception.[71] The reasons for their endorsement of the Capper-McFadden measure were not clear. Whether the generally conservative livestock men chose to follow Mellon's advice or whether they felt that they could better handle their own credit needs through the private corporations which would be le-

[69] Wallace, Hoover, and Mellon gave their views before closed hearings of the Senate Committee on Banking in December 1922. *Washington Post,* December 31, 1922, p. 4.

[70] Peter Norbeck to P. W. Doughtery, December 16, 1922; Norbeck to S. W. Clark, January 29, 1924, Norbeck Letters, Box 1; and *Washington Post,* December 13, 1922, p. 8.

[71] *Washington Post,* December 31, 1922, p. 4.

galized under the measure, the fact remained that the nine-month discount period provided for in the bill would have been clearly inadequate for their needs.

Harding had little interest in either bill except insofar as it would fulfill his campaign promise for improvement of rural credit facilities. At the opening of Congress in December 1922, Wallace wrote to the President that "the most urgent need now so far as legislation is concerned is a credit system better adapted to the needs of the farmer and stock grower," and he urged Harding to recommend passage of the Lenroot-Anderson bill.[72] As the two credit measures went to committee, however, the President remained silent. Early in 1923 both bills were reported out and passed by the Senate, but disagreement in the Committee on Banking delayed House action on them. Complaining that the committee displayed "no particular enthusiasm" to act on the Lenroot-Anderson bill, Wallace advised Harding that it would be unfortunate "if the . . . bill should not become a law" and suggested that if he felt justified in "letting some of the people on the committee know of your interest, it would be most helpful." [73] Reluctant to become involved in the congressional fight, Harding refused to back either bill and the deadlock in the House continued.

In late February differences among cabinet members came fully into the open, a development which helped to bring about a resolution of the issue over agricultural credit. Replying to a request from McFadden for his views on the two credit bills before the House, Mellon wrote a long letter which was released to the press. Maintaining that the Lenroot-Anderson measure was ill-conceived, the Treasury secretary charged that the proposal violated "every canon of sound banking to which this Government

[72] Wallace to Harding, December 3, 1922, Files of the Secretary of Agriculture.
[73] Wallace to Harding, February 12, 1923, Harding Papers, Box 1.

has been committed since the establishment of the national banking system." The Capper-McFadden proposal, on the other hand, was sound and, even more important, "avoids the excessive centralization which . . . constitutes a serious defect in the Lenroot bill." [74]

Sydney Anderson was alarmed lest the published views of Mellon be taken as those of the administration in general. He therefore requested that Wallace and Hoover restate their positions on the credit bills "in such a way that no doubt may exist either in the minds of Members of Congress or the public." [75] In his statement, which was also released to the newspapers, Wallace asserted that the Lenroot-Anderson bill was a "true rural credits measure" because it provided for the supervision of the farmers' money needs by an existing federal agency already familiar with agricultural problems. On the other hand, Capper's proposal could make no such claim for it was "designed to encourage by Government authority the organization of private corporations organized and operated for the profit of their stockholders. . . ." Though Wallace had no objection to passage of the Capper-McFadden bill, he explained that to offer it as a substitute for the Lenroot measure "would give farmers of the Nation the best of reasons for feeling that nothing was done." In his published reply to Anderson, Hoover agreed with Wallace and urged enactment of the Lenroot-Anderson bill at the earliest possible moment.[76]

With no sign of a break in the House deadlock, Harding finally intervened. He petitioned Republican members of the House Committee on Banking to work out a compromise which would be acceptable to both factions so that the party's campaign prom-

[74] Mellon to McFadden, February 19, 1923, copy in Files of the Secretary of Agriculture.

[75] Anderson to Wallace, February 22, 1923, *ibid.* Hoover received the same letter.

[76] Wallace to Anderson, February 23, 1923, *ibid.;* and Hoover to Anderson, February 23, 1923, Hoover Papers, 1-I/4.

ise on rural credits might be fulfilled before adjournment of the Sixty-seventh Congress.[77] Moving rapidly, the committee reached agreement on a bill incorporating the principal features of both of the measures under consideration. Endorsed by the President, the bill quickly passed Congress and was signed into law as the Intermediate Credits Act in early March 1923.

The act established 12 intermediate credit banks in districts corresponding to those of the Federal Reserve System, but under a separate board established specifically for the administration of the new system. They were empowered to discount the agricultural paper of national and state banks and farm credit cooperatives, and to make direct loans to marketing associations for periods of up to three years. Initially capitalized by the federal government, the banks were authorized to issue tax-exempt debentures once operations had begun. As a concession to backers of the Capper-McFadden bill, the act also provided for the incorporation under federal charter of private corporations to supply credit directly to farmers.

Wallace was well pleased with the compromise measure, and he praised Harding's "vigorous action" in its behalf. In a press release immediately after enactment of the Intermediate Credits Act, he cited it as one of the most important accomplishments of the Sixty-seventh Congress. Despite the fact that the act was designed specifically to meet the credit needs of the country's farmers, the secretary took exception to charges that it was merely a piece of class legislation. "The results," he claimed, "will be helpful to business in general, because the effect will be to stabilize agricultural production and marketing." [78] Emphasizing that the Intermediate Credits Act was not an emergency relief measure,

[77] *Washington Post,* February 24, 1923, p. 4.
[78] USDA press release, March 6, 1923, copy in files of the Secretary of Agriculture; *Washington Post,* March 20, 1923, p. 5; and *Smoky Mountain Husbandman,* December 6, 1923.

Wallace viewed it as the establishment of a permanent system of credits adapted to the long-term needs of farmers. By supplying a source of loans designed to aid in the orderly market of farm products, Wallace pointed out, the act would help agriculture to place its operations on a more businesslike and rational basis.[79]

Wallace also took an active interest in the controversy over the government-owned hydroelectric and nitrate plants at Muscle Shoals, Alabama, though he did not become involved in the congressional fight over it. The Wilson administration had constructed the works during World War I to provide nitrates for military purposes. Anxious to dispose of the plants, Harding was favorable toward an offer by Henry Ford to purchase the property and urged Congress to pass the necessary legislation. Although Wallace supported the sale to Ford, he emphasized that the terms of the deal should be sufficiently binding to insure the production of cheap nitrates for use in fertilizer. At the same time the secretary sent recommendations to the Senate for government operation of the nitrate plants in the event that the property was retained.[80] Under the leadership of Senator George W. Norris, public power advocates blocked passage of the Muscle Shoals bill and Ford finally withdrew his offer.

Wallace's efforts in behalf of the farm legislation passed in his first two years in office illustrate well his ideas on the function and responsibility of the Agriculture Department. In addition to being an administrative and policy-making agency, the department was also to take an active part in the legislative process. Not content with merely advising lawmakers, Wallace used his

[79] Wallace to A. E. DeRicqles, November 16, 1923, Files of the Secretary of Agriculture.

[80] "Official Statement of the Secretary of Agriculture on Muscle Shoals," February 14, 1922, ms.; Wallace to George W. Norris, May 23, 1923, Files of the Secretary of Agriculture; and Wallace to Coolidge, Calvin Coolidge Papers, Manuscript Division, Library of Congress, File 44, Box 67.

office to bring pressure upon Congress for measures he believed farmers needed. Although he developed close relationships with few on Capitol Hill, legislators generally recognized his superior knowledge of agriculture and respected his opinions. Furthermore, the secretary enjoyed a warm friendship with Harding. The two men often played cards and golf together, and Wallace used these moments of diversion as well as official channels of communication to encourage the President's endorsement of desired bills. As a result, he exercised considerable influence in Congress and assumed a larger role in the enactment of laws than is usually associated with cabinet members.

At first Wallace was well satisfied with the accomplishments of the Sixty-seventh Congress. He believed that the farm legislation it had enacted represented an important step in meeting the immediate needs of farmers, as well as a sound beginning of a permanent federal agricultural program. All of the measures, furthermore, called for limited government intervention in the economy and thus conformed to his concept of federal responsibility. When the expected improvement in farm conditions failed to materialize, however, Wallace changed his mind. "These laws," he decided, "were good, but were rather a treatment of symptoms than of the disease." As he came to realize, the principal weakness of the legislation was its failure to deal effectively with the basic cause of the farm depression of the 1920's—continuing surpluses.[81] In the last year before his death in 1924, therefore, he would expand his concept of government responsibility and endorse a more far-reaching solution to the agricultural crisis.

[81] Wallace, *Our Debt and Duty to the Farmer,* pp. 102–3; and "Report of the Secretary," USDA *Yearbook of Agriculture,* 1924, p. 24.

Chapter VI / A PROGRAM
OF AGRICULTURAL ECONOMICS

ALTHOUGH Secretary Wallace supported the work of farm representatives in Congress and devoted a good deal of his own time and effort to the promotion of their program, he did not view the legislation of the early 1920's as an agricultural panacea. He firmly believed that a fundamental modification was necessary within the operation of the farming industry itself before producers would be able to earn a fair return on their labor and investment, indeed, before they could hope to reap full benefit from a federal agricultural program. Under Wallace the Department of Agriculture became the vehicle for that change, a change which, despite the fact that it fell short of his expectations, in many respects represented the secretary's most significant contribution to the development of a national farm policy.

Wallace's solution to what he saw as the major weakness of agriculture grew naturally out of his analysis of American development. To the secretary, the United States was an arena of conflicting economic interests, with agriculture, labor, and capital as the major antagonists. Throughout the country's history there had been a continuing contest among these interests over the division of national wealth and income, a struggle in which the farming sector had fallen steadily behind. Agriculture's failure to maintain its place was not difficult to explain. Members of the capital and labor groups had acquired a thorough understanding of the American economy, which enabled them to anticipate and adjust to changing economic conditions.[1] But farmers, having failed to keep pace in learning to function within the competitive system, found themselves without a workable strategy for coping with its problems and at a marked disadvantage in the contest with their more skilled adversaries.[2]

Although Wallace often criticized the methods of capital and labor, he nevertheless encouraged farmers to follow their example. If agriculture were to regain its proper place in the economy, he urged as an editor, producers of food and fiber would have to develop effective techniques for the protection of their interests. To accomplish this required a thorough grasp of the concept of supply and demand and of the price-making process. "The notion," he admonished, "that farmers can have anything to say about prices of farm products until [sic] they have gotten a real understanding of the forces which make prices is an iridescent dream." They had to have the knowledge and information which would enable them to adjust their production to existing and anticipated demand, market their products in orderly fashion, and rationalize their operations. Only then would they be able to influence prices in the same way as other groups and to function effectively within the competitive system.[3]

In 1918 Wallace ran a series of editorials in *Wallaces' Farmer* proposing the creation of a farmers' university. Supported and

[1] *Wallaces' Farmer* 43 (September 27, 1918): 1368; 45 (January 23, 1920): 258; and 45 (November 5, 1920): 2559.

[2] *Ibid.*, 44 (February 28, 1919): 538 and 45 (October 8, 1920): 2355.

[3] *Ibid.*, 44 (February 28, 1919): 538; 45 (November 5, 1920): 2559; 45 (December 10, 1920): 2748; Wallace to Charles H. Betts, June 30, 1921, Files of the Secretary of Agriculture, National Archives; and address before the Farmers' Grain Marketing Committee, Chicago, April 16, 1921, copy in Henry C. Taylor Papers, State Historical Society of Wisconsin, Madison.

controlled entirely by farmers, the institution was to educate a corps of agricultural economists who would go out and train producers in the methods and techniques of sound business practice. They would teach them to keep account books and calculate production costs, to interpret market reports and price trends, to anticipate and adjust to fluctuations in the economy, and to apply principles of systematic marketing and distribution. In addition, they were to represent farmers in Washington as congressional lobbyists and before the various federal regulatory agencies. Equipped with an understanding of foreign and domestic economy and indoctrinated with an agrarian viewpoint, these representatives would be prepared to defend agriculture against encroachment from other sectors of the economy.[4]

If the scheme was impractical, Wallace, at least at the time, saw it as the only hope for a program of economic service to farmers. Of the opinion that there were several agencies which could and should accept this responsibility, he regretted that none had shown more than incidental inclination to do so. In his editorials he attacked the agricultural colleges for failing to take the lead in studying the business side of the farm situation. He charged that the curricula in the schools focused instead on improvements in the techniques of production for the purpose of teaching farmers how to produce larger crops and better livestock at a lower cost, thereby insuring the country a cheap food supply.[5] Nor was the Department of Agriculture under the current Woodrow Wilson administration, to Wallace's thinking, any more effective in attacking the farmers' economic problems. Even the data gathered by the Bureaus of Crop Estimates and Markets were prepared from the standpoint of encouraging

[4] *Wallaces' Farmer* 43 (September 27, 1918): 1368 and 43 (November 15, 1918): 1677.

[5] *Ibid.*, 43 (November 15, 1918): 1677 and 44 (March 21, 1919): 716.

greater production rather than with the idea of supplying information which would enable farmers to bargain more effectively.[6] In contrast, Wallace pointed out, the Labor Department supplied workers with accurate statistics on the cost of living, wages, strikes, and general employment conditions and attempted to develop among laborers an appreciation of their strategic position within the economy. Secretary of Agriculture David F. Houston, however, lacked a sympathetic understanding of the problems of farmers and was therefore disinclined to develop his department along similar lines.[7]

Wallace likewise considered much of the activity of agricultural organizations misdirected. Charging that they often dissipated their energies on futile political agitation and gave little thought to the business problems of agriculture, he reminded the organizations that until producers attained greater economic power they could not expect to command any significant political power. Since farm problems were primarily of an economic nature, furthermore, most political remedies would at best bring only temporary relief. Even the marketing associations did not escape Wallace's criticism. Too often, he claimed, they concentrated entirely on the techniques of creating and administering the organization itself and neglected to apply the more important economic principles upon which cooperation must be based if it were to succeed.[8]

Wallace pointed to the Corn Belt Meat Producers' Association as a model of the kind of organization that could be of real service to producers. The association was successful because "it has

[6] Ibid., 43 (November 29, 1918): 1741 and 43 (December 13, 1918): 1812.

[7] Ibid., 43 (September 27, 1918): 1368; 44 (January 3, 1919): 5; and 45 (October 8, 1920): 2355.

[8] Ibid., 43 (September 13, 1918): 1289; 45 (April 16, 1920): 1143; 45 (August 13, 1920): 1946; and Wallace to Gifford Pinchot, September 10, 1917, Henry C. Wallace Papers, University of Iowa Library, Iowa City, Box 1.

not depended upon resolutions and committees, but upon facts and figures." [9] As editor of *Wallaces' Farmer,* he suggested creation of a "Chamber of Agriculture," patterned after the National Chamber of Commerce, the function of which would be to study and interpret the economic situation for the purpose of advising farmers.[10] Instrumental in the formation of the American Farm Bureau Federation, Wallace urged the organization to adopt a similar program. "This federation," he cautioned, "must not degenerate into an educational or social institution, . . . [but] must be made the most powerful business institution in the country." [11] As guest speaker at the final organizational meeting in March 1920, he warned that, "if the purpose of this organization is to carry on the sort of work which the Farm Bureaus have been doing heretofore, which is for the purpose of education, for the purpose of stimulating production, . . . then the Farm Bureau organization as you have started it now will serve no useful purpose; in fact, it will do harm." [12] The Farm Bureau's failure to carry through his plan was a disappointment to Wallace.[13]

Wallace's attitude toward agricultural economics was the principal factor in his decision to accept the post of Secretary of Agriculture in the Harding administration. The precipitous decline in prices in 1920 had further impressed upon him the need for the program he had been urging. The crisis, Wallace claimed, was due in large part to the fact that "no one representing the

[9] *Wallaces' Farmer* 45 (February 27, 1920): 690.

[10] *Ibid.,* 45 (February 7, 1919): 589. Wallace's suggestion was discussed at a meeting of Agriculture Department heads before he became secretary, but no action was taken. "Minutes of bureau chiefs' meeting, May 6, 1920," Taylor Papers.

[11] Quoted in Orville M. Kile, *The Farm Bureau through Three Decades* (Baltimore: Waverly Press, 1948), pp. 56–57; and *Wallaces' Farmer* 44 (October 24, 1919): 2127.

[12] Quoted in Christiana McFadyen Campbell, *The Farm Bureau and the New Deal* (Urbana: University of Illinois Press, 1962), p. 31.

[13] *Wallaces' Farmer* 45 (April 16, 1920): 1143.

farmers has a thoro understanding of the forces which are hitting them or how to combat these forces." If producers were going to be able to protect their interests during the turbulent postwar period, now more than ever they would have to be made aware of the business side of farming and supplied with accurate information. Since no agency had taken upon itself this task, Wallace decided to accept the position as Secretary of Agriculture and develop the department along economic lines.[14]

Despite Wallace's denunciation of the Agriculture Department, progress had been made toward establishment of a farm economics program during the Wilson administration. Directing his attention toward what he called the "other great half" of the agricultural question, Secretary Houston had devoted considerable attention to the development of economic research and services. For the purpose of gathering and disseminating information on the marketing and distribution of farm products, he established in 1913 the Office of Markets, which became the Bureau of Markets four years later. The need for greater efficiency and coordination in the farming industry to meet the country's agricultural needs during World War I resulted in further expansion of the department's economic activity. After the war Houston consolidated much of the work being done in this area under a new agency called the Office of Farm Management and Farm Economics. His successor, Edwin T. Meredith, continued the development of economic activities after he took over as secretary in 1920.[15]

Envisioning a somewhat more extensive program than the one initiated under the Wilson administration, Wallace undertook to

[14] *Ibid.*, 45 (October 8, 1920): 2354 and 45 (November 5, 1920): 2559.
[15] Henry C. Taylor, "A Farm Economist in Washington," pp. 43–44, 135–36, ms.; Taylor to E. T. Meredith, July 23, 1921, Taylor Papers; and David F. Houston, *Eight Years with Wilson's Cabinet* (New York: Doubleday, Page, 1926), vol. 1, pp. 199–200.

prepare the groundwork for several changes soon after taking office. His first action was the appointment of a committee to study the department's economic work and to make recommendations for reorganization and expansion.[16] Reporting in June 1921, the committee proposed an increase in economic research and service and the creation of a new bureau to direct the work along these lines. When the proposal met with the approval of most of the department's bureau chiefs and received strong endorsement from farm organizations, Wallace moved rapidly to implement the committee's recommendations.[17]

The secretary placed Henry C. Taylor in charge of working out a plan of reorganization. One of the country's leading agricultural economists, Taylor had been interested in the development and practical application of this relatively new social science for some time. As a professor at the University of Wisconsin from 1901 to 1919, he was instrumental in bringing about an expansion of teaching and research in farm economics within the College of Agriculture. What little development had been made in this area at other schools was directed mainly toward the improvement of management and the rationalization of operations on the individual farm. Under the influence of Taylor, Wisconsin took a more comprehensive approach and concentrated to a greater degree upon the questions of agriculture's relationship to the economy in general and improvements in the industry as a whole.[18] Houston brought the Wisconsin professor to

[16] Taylor to Chester C. Davis, January 25, 1926, Taylor Papers; and Taylor to F. A. Pearson, May 28, 1921, Files of the Bureau of Agriculture Economics, National Archives.

[17] Note dated July 16, 1921, Files of the Secretary of Agriculture; and Henry C. Taylor and Anne D. Taylor, *The Story of Agricultural Economics, 1840–1932* (Ames: Iowa State College Press, 1952), p. 602.

[18] H. L. Russell, "Comments on the University Survey Report of the College of Agriculture," December 18, 1914, ms. in files of the College of Agriculture of the University of Wisconsin, University of Wisconsin Archives, Madison, Series 9/1/2, Box 24.

Washington in 1919 to head the newly created Office of Farm Management and Farm Economics, and in that position he did much to direct the department's attention toward the field of agricultural economics. Wallace and Taylor, therefore, shared a wide area of agreement on how to meet the needs of American farmers.

The first step in reorganization took place on July 1, 1921, with the merger of the Bureaus of Markets and Crop Estimates, an action previously authorized by Congress upon the recommendation of Meredith in the fall of 1920.[19] Also called for in the study committee's plan was the incorporation of the Office of Farm Management and Farm Economics within the new division, but this had to await congressional approval. In the meantime, Taylor coordinated and redirected the activities of both offices in such a way as to effect virtual consolidation before it was actually authorized. Formation of the Bureau of Agricultural Economics (BAE) on July 1, 1922, incorporating all economic functions of the department into a single division, completed the reorganization program. Wallace appointed Taylor as chief of the new division, which soon became the most active bureau of the department and one of the outstanding research agencies of the federal government.[20]

Wallace viewed the creation of the BAE as the major accomplishment of his first year in office, going so far as to say that it marked the beginning of the end of the period of agricultural exploitation. Regretting that similar action had not been taken sooner, he suggested in his first report that "had we in the past given as much attention to the economics of agriculture as we have to stimulating production . . . some of the troubles which

[19] Memorandum, Leon M. Estabrook to Wallace, March 16, 1921, BAE Files.

[20] "Report of the Secretary," USDA *Yearbook of Agriculture*, 1921, pp. 16–17; and USDA *Weekly News Letter*, August 31, 1921, p. 5.

now beset us might have been anticipated and avoided." Not only would the department's program of farm economics prove useful for immediate relief, Wallace could boast, but, of greater long-term significance, it held out the promise of a rejuvenated agricultural industry.[21]

As the BAE developed, it came to offer a wide range of services. The new division gathered and interpreted crop and livestock statistics, studied various aspects of marketing and distribution, established and enforced commodity standards, provided information aimed at improved farm organization and management, and investigated foreign production and demand. Under its direction extension work in farm economics was expanded, the market news service improved, and publication of agricultural outlook reports initiated. To carry out its work the BAE required a large staff of experts, and the bureau attempted to meet this need by establishing an extensive in-service training program.[22]

The BAE concentrated a good deal of attention on and applied considerable effort toward the improvement in methods of data-gathering, reporting, and forecasting.[23] Since its formation in 1862, the Department of Agriculture had been issuing semi-monthly reports on crop production based on information received from volunteer compilers in each county of the country.

[21] Wallace to Charles H. Betts, May 30, 1921; Wallace to Wealy L. Jones, March 20, 1922, Files of the Secretary of Agriculture; and "Report of the Secretary," USDA *Yearbook of Agriculture,* 1921, pp. 16–17.

[22] USDA *Weekly News Letter,* August 31, 1921, pp. 5–6; Taylor, "A Farm Economist in Washington," p. 238; and Taylor, *The Story of Agricultural Economics,* pp. 710–11. Discussion of the foreign work of the BAE will be left to a later chapter.

[23] Henry C. Wallace, *Our Debt and Duty to the Farmer* (New York: Century, 1925), p. 153; Henry C. Taylor, "A Century of Agricultural Statistics," *Journal of Farm Economics* 21 (November, 1939): 703; Wallace to H. A. Wallace, September 8, 1922, Files of the Secretary of Agriculture; and "Report to Wallace on Work of the Department," September 29, 1922, Taylor Papers.

In 1867 data on current price levels for various commodities were added to these publications, and the first official forecasts of production prior to harvest appeared in 1912. Not until 1917, however, did the department begin to take preplanting surveys on the plans of farmers regarding how much land they intended to devote to various crops. And since these were incomplete experimental studies, their results were withheld from publication and thus were of no help to producers. Despite some interest along these lines prior to 1921, the Agriculture Department had given relatively little time or money to improving and expanding the methods of obtaining data or to refining the interpreting and reporting of statistics.[24]

In the first significant expansion of reporting and forecasting work under Wallace, the department published its initial pig report in June 1922. Earlier in the year the Bureau of Markets and Crop Estimates had prepared an "intentions-to-breed" questionnaire, which rural mail carriers delivered to farmers and, upon completion, returned to the Agriculture Department. The department asked hog producers to supply information on the size of their droves, the number of sows, and their plans for fall farrowing. After compiling and interpreting the data, the bureau published its findings in the June pig report. According to department figures based on the survey, farmers intended to increase hog production by about 48 per cent in 1922, but the actual increase amounted to only 28 per cent. Immensely pleased, Taylor attributed the difference to the wide publicity given to the June report, explaining that after learning of the anticipated rise in hog production, farmers cut back on their original plans for fall farrowing.[25] Viewing the results of the pig report as indica-

[24] Walter H. Ebling, "Why the Government Entered the Field of Crop Reporting and Forecasting," *Journal of Farm Economics* 21 (November, 1939): 724–30.

[25] Memorandum, Taylor to Wallace, April 27, 1923, Taylor Papers; and Taylor, "A Farm Economist in Washington," pp. 113–14.

tive of the kind of adjustment agriculture might make if supplied with the necessary information, Wallace and Taylor enthusiastically prepared to apply the "intentions" survey method to other commodities.

The Department of Agriculture followed up its pig report with "intentions-to-plant" surveys on various crops. Again rural mail carriers, as well as BAE representatives in the field, gathered the necessary information. Once the data had been compiled, Wallace called together a committee of economists and farm leaders from outside of the government, representatives of the Agriculture and Commerce Departments, and members of the Federal Reserve Board to analyze the material and forecast trends in production and demand. Meeting on April 20 and 21, 1923, the Agricultural Outlook Conference published its findings along with the intentions-to-plant surveys of the BAE a few days later in the first Agricultural Outlook Report.[26] Wallace assembled the group for another conference on July 11 and 12 to deal specifically with the questions of foreign demand for American farm goods and of wheat, corn, and hog production. Using the most recent intentions-to-plant and -breed surveys of the BAE, the second meeting also made public its forecasts on production and demand.[27] Beginning in March 1924 the outlook conference and published report became an annual occurrence. Due to criticism of employing outside economists and farm representatives, however, subsequent conferences used only department experts.[28]

Designed as they were to make available information on future supply and demand, the Agricultural Outlook Reports were

[26] *Agricultural Situation* (USDA), May, 1923, p. 13; Taylor, "A Farm Economist in Washington," pp. 136–37. Henry A. Wallace, upon Taylor's invitation, was one of the members of the conference. Taylor to H. A. Wallace, April 6, 1923; H. A. Wallace to Taylor, April 26, 1923, Taylor Papers.

[27] *New York Times*, June 25, 1923, p. 26.

[28] *Iowa Union Farmer*, May 30, 1923, p. 4; and Taylor, "A Farm Economist in Washington," pp. 138–39.

expected in some mysterious way to induce farmers to adjust their production accordingly. Confident that producers would individually increase or decrease output of specific commodities in light of the projected conditions forecasted in the reports, Wallace considered publication of these data as "one of the best measures we could have to equalize production." [29] The Agriculture Department made no effort to advise farmers on what they should or should not produce or how they might put the outlook data to use. As Taylor explained it, the purpose was only to supply "the facts they [farmers] needed in order to act intelligently." [30] Once provided with the basic information, so the thinking went, producers could be depended upon to make the necessary adjustments. Optimism was indeed an important element in the department's program of farm economics.

While most agricultural interests welcomed the department's crop and livestock reporting and forecasting, southern cotton planters were from the beginning resolutely opposed. Under the leadership of the influential American Cotton Association, planters fought the inclusion of cotton data in the semimonthly reports and annual outlook publications. Having established an extensive cotton reporting service of its own, the association saw no need for the Agriculture Department to duplicate the work. Planters felt, moreover, that their own service was superior to that of the BAE, a notion which gross inaccuracies in the department's 1921 cotton estimates did little to dissuade.[31] To aggravate matters further, the first Agricultural Outlook Report of

[29] United States Congress, House, *Hearings on the McNary-Haugen Export Bill,* 68th Cong., 1st Sess., p. 126; and "Report of the Secretary," USDA *Yearbook of Agriculture,* 1923, p. 20.

[30] Taylor, "A Farm Economist in Washington," p. 137.

[31] J. S. Wannamaker to Leon M. Estabrook, June 24, 1922; W. F. Callander to Taylor, January 4, 1923, Files of the Secretary of Agriculture.

April 1923 forecast a 12 per cent increase in cotton acreage for the following year and estimated a large carry-over for the current year, publication of which caused a sharp drop in prices. If this were not enough, a month later the BAE had to admit that it had overestimated the 1923 carry over.[32]

Understandably incensed, cotton planters rallied behind J. S. Wannamaker, president of the American Cotton Association, in a campaign to force the BAE to drop its cotton forecasts. In letters to association members, Wannamaker urged them to protest to their congressmen "against interference in price mechanism" by the Agriculture Department.[33] Meeting with Wallace in December 1923, a committee representing the organization demanded an end to intentions-to-plant reports on cotton and the appointment of cotton farmers to the department's production estimating board. Early 1924 saw the introduction of legislation into Congress which, among other things, proposed to prevent the department from publishing agricultural forecasts of any kind. Alarmed by the flood of protests and the action in Congress, Taylor met with southern congressmen, who were supporting the measure, and agreed to eliminate cotton estimates from the 1924 outlook report in return for a promise to drop from the bill the section prohibiting forecasting.[34] Announcement of the compromise was followed by a decline in interest in the bill, with the result that it was never brought to a vote. Despite the fact that the prohibitory section was not deleted from the proposed legislation, the department apparently decided that discretion

[32] J. S. Wannamaker to Wallace, May 21, 1923; Taylor to Wallace, May 31, 1923, *ibid.*, and *New York Times,* May 6, 1923, p. 6.

[33] J. S. Wannamaker to Jewell Mayes, April 28, 1923; and Wannamaker to Wallace, May 7, 1923, Files of the Secretary of Agriculture.

[34] Wallace to James F. Byrnes, February 4, 1924; Wallace to William T. Harris, February 15, 1924, *ibid;* and *Commonwealth Appeal,* December 22, 1923.

was in order and omitted cotton data from its 1924 outlook report anyway.[35]

Criticism of another kind came from the Farmers' Union. Milo Reno, president of the Iowa organization, charged that the Agricultural Outlook Conference of April 1923 had been dominated by interests hostile to farmers. "The danger to Agriculture," he warned, "is now, as it always has been, that those who are interested in exploiting agriculture should presume to take a leading part in determining agricultural affairs." [36] Although there were representatives of bankers and meat packers at the conference, to say that they dominated the meeting was an exaggeration. Also present were officers of agricultural organizations as well as economists from several of the country's leading colleges and universities. Henry A. Wallace, for instance, took a major part in the conference, and he was hardly antagonistic toward farm interests. Suspicious of any solution to agricultural problems short of government price-fixing or federal ownership, the Farmers' Union frequently attacked Wallace's department for its more cautious approach.

As agricultural conditions showed few signs of improvement, the original enthusiasm of farmers for crop and livestock reporting began to wane. "Your old statistics are an awful help to the speculator," wrote a distressed producer to Wallace, "but a terrible hurt to the farmer." Another was amused at the department's "assumption that it can get and publish what is in the farmer's mind." [37] In many respects these assessments were valid. Besides the difficulties involved with gathering accurate data, publication of forecasts tended to induce producers—in fact, this was the

[35] Wallace to J. B. Aswell, February 23, 1924; Wallace to C. A. Cobb, February 29, 1924; and Wallace to James F. Byrnes, April 3, 1924, Files of the Secretary of Agriculture.

[36] *Iowa Union Farmer*, May 30, 1923, p. 4.

[37] N. D. Crosly to Wallace, August 2, 1922; and Theodore H. Price to Wallace, January 18, 1924, Files of the Secretary of Agriculture.

main purpose—to react in such a way as to reduce the validity of the predictions. As a result, the BAE was constantly open to charges of inaccurate and misleading reporting. Contrary to expectations, moreover, farmers, acting individually, were incapable of making intelligent use of the information provided by the BAE, and agriculture lacked the kind of organizational machinery necessary for the planning and adjustment envisioned by the Agriculture Department.

When reporting and forecasting failed to bring about the elimination of surpluses and the desired balance in production and demand, members of the department itself began to doubt the utility of their program. Disillusionment with voluntary adjustment would be an important factor in the department's later endorsement of more aggressive measures for dealing with the problem of domestic surplus. Agricultural forecasting nevertheless remained an important part of the work of the BAE.

If reporting and forecasting were designed to aid in bringing about balanced production, the department's work with cooperatives was aimed at improved methods of marketing and distribution. Wallace had long felt that cooperative associations, if properly administered, offered one of the best solutions to the problems of agricultural marketing, and he predicted that eventually "farmers will market most of their products through cooperative associations, thereby eliminating much of the unreasonable toll now collected by middlemen." [38] Thus the secretary believed that it was as much the responsibility of his department to help farmers realize this goal as it was to aid them in the adjustment of their production.[39]

When Wallace took office the Department of Agriculture

[38] *Wallaces' Farmer* 45 (August 13, 1920): 1946; and Wallace, *Our Debt and Duty to the Farmer*, p. 155.

[39] Address before the Farmers' Grain Marketing Committee, Chicago, April 6, 1921, copy in Taylor Papers.

lacked a definite policy regarding its relationship to cooperative enterprises, the result of which was a good deal of misunderstanding as to the type of departmental help these organizations could expect. Speaking before the Farmers' Grain Marketing Committee soon after taking office, the secretary sought to clarify his position on this question by announcing the guidelines he intended to follow. Wallace explained that the government could properly collect, compile, and analyze the essential data pertaining to marketing associations, study the economic principles upon which successful cooperation must be based, and disseminate this information among farmers and their organizations. But he emphasized that "there are limits beyond which the department cannot properly pass in the field of organization of cooperative associations." Clearly, the secretary insisted, the limits of federal responsibility and power did not permit the Agriculture Department, as some were urging, to take an active part in the initiation, formation, and management of associations or to give official endorsement to any particular group. To be truly cooperative, Wallace held, organizations had to remain entirely free of government control, domination, or direction.[40] Consistent with his views on production adjustment, the secretary believed that his office should supply cooperative information, after which it became the farmers' responsibility to utilize that information, individually and collectively, in furthering their own interests.

The Agriculture Department created a special division within the BAE to handle research and service in the area of cooperation. This agency investigated the operations of unsuccessful as well as successful organizations, evaluated various marketing, distribution, and management procedures, studied improved sys-

[40] Wallace to George Livingston, March 28, 1921, BAE Files; address before the Farmers' Grain Marketing Committee, Chicago, April 6, 1921, copy in Taylor Papers; *New York Times,* March 9, 1924, p. 8; and "Report of the Secretary," USDA *Yearbook of Agriculture,* 1923, p. 35.

tems of accounting, and analyzed economic conditions as they affected cooperative activity. Serving as a source library on the subject of cooperation, the BAE made its information readily available to any existing or proposed associations. Presented in special publications and in the department's yearbooks, its findings represented the most complete study of cooperative activity available to farmers.[41] The bureau also provided advisors for groups interested in organizing cooperatives.[42]

There were those, of course, who criticized the department for not doing enough to promote cooperative activity. Some wanted the government to take the lead in forming associations, while others urged federal control and operation. Beginning in 1922 legislation to broaden federal participation in the cooperative movement was regularly introduced into Congress, but it was rejected until enactment in 1926 of the Cooperative Marketing Act, providing for a modest extension of the duties of the BAE in this area. Not until the passage of the Agricultural Marketing Act of 1929, under which the Federal Farm Board took over responsibility for cooperative activity from the Agriculture Department, did the work of the government go beyond research and advising. Despite increasing pressure for a more aggressive policy as farm conditions remained depressed, Wallace continued resolute in his opposition to greater federal involvement in the field of cooperation until his death in 1924. To his brother he wrote that "the Department is thoroughly committed to helpful activity with regard to cooperative associations. We have a vast amount of material here and several men who are real authorities. Much

[41] "Report of the Secretary," USDA *Yearbook of Agriculture*, 1923, pp. 42–43. For examples of studies on different types of cooperatives see the ones on dairy and tobacco associations in the 1922 *Yearbook of Agriculture* and those on egg and poultry associations in the 1924 *Yearbook of Agriculture*.

[42] Taylor to Wallace, September 19, 1923, and attached report, Files of the Secretary of Agriculture.

criticism has come because we have not turned in and boosted some particular form of cooperative enterprise." The strength of the movement, he told Senator George W. Norris, "indicates that there is no general need for a departure from the present methods of organizing such associations." [43]

In spite of a marked growth in the number of cooperatives in the early 1920's, by 1923 it was evident that orderly marketing and distribution through agricultural associations were contributing little to solving farm problems. Some gains were made in the areas of balanced distribution, movement of goods to market, grading and standardization, advertising, and increased bargaining power in dealings with buyers. But no solution appeared to the problem of continuing surpluses, which, even with the most rational and systematic methods of marketing and distribution, would prevent improvement in farm conditions.[44]

In addition to its work directly with cooperatives, the BAE also improved and expanded the department's market news service in an effort to aid farmers in the marketing of their products. For some time the government had distributed information on prices and receipts at the major buying centers, but the service had proved to be inadequate. Since newspapers were the only source of Agriculture Department reports available to most farmers, they received the information one or two days late. In many cases, therefore, producers had to take their goods to market without knowing what they would bring in the way of prices. It was important, Wallace thought, for farmers and cooperatives to learn as soon as possible the prices and anticipated receipts at the

[43] Wallace to Dan Wallace, December 22, 1923, Wallace Papers, Box 2; and Wallace to Norris, April 12, 1922, Files of the Secretary of Agriculture.

[44] James H. Shideler, "Herbert Hoover and the Federal Farm Board Project, 1921–1925," *Mississippi Valley Historical Review* 42 (March, 1956): 713.

buying centers so that they could market their products to greater advantage.[45]

During Wallace's tenure as Secretary of Agriculture, the most significant improvement in the market news service was in the expanded use of radio for the dissemination of information. When he assumed office radio transmission of prices and receipts was relatively new and employed only to a very limited degree by the department. After taking over the news service in 1922, the BAE conducted a study on the possibility of wider use of the radio for this purpose and found that it would be feasible from the department's standpoint and helpful to farmers. Under an arrangement with the Navy to borrow its broadcasting facilities, the BAE relayed market reports to state agriculture departments, extension services, and agricultural colleges, which in turn passed on the material to farmers in their areas either by rebroadcast over local stations or printed report.[46] Although lack of funds prevented the fullest use of radio transmission, the market news service was made much more effective through the limited expansion along these lines. It gave a larger number of farmers ready access to important information on prices and receipts and thus enabled them to market their products more intelligently.[47]

The BAE also attempted to rationalize the marketing of agricultural products through the establishment of uniform standards. With the grades of farm commodities in doubt, buyers were inclined to discount prices to protect themselves against in-

[45] Wallace to James W. Good, April 25, 1921, Files of the Secretary of Agriculture.

[46] George Livingston to Wallace, May 25, 1921; Lloyd S. Tenny to Wallace, March 8, 1923, ibid.; J. Clyde Marquis to Taylor, February 29, 1924; C. W. Kitchen to Taylor, February 29, 1924, Taylor Papers; and Taylor, The Story of Agricultural Economics, pp. 633–35.

[47] Wallace to F. G. Ketner, April 19, 1924, Files of the Secretary of Agriculture.

ferior quality, a practice which of course worked to the disadvantage of producers of high quality goods. Absence of uniform standards and grading, furthermore, led to confusion in interpreting price quotations, thus enabling speculators to manipulate the market and deceive sellers. If official grades could be established, the BAE reasoned, farmers would receive prices more reflective of the quality of their products, would be less vulnerable to the manipulations of speculators, and, along with others in the trade, would enjoy the benefits of a more efficient marketing system.

Under Wallace the department's most important work in this area was the establishment of a set of worldwide cotton grades based upon American standards. The Cotton Futures Act of 1914, administered by the Agriculture Department, required the use of federal grades in futures transactions on the domestic market. Early efforts to gain international adoption of American standards, however, had been blocked by the Liverpool Cotton Association of England. With its own set of grades, which was generally recognized in international trade, the association was understandably reluctant to accept a substitute. American planters and exporters, on the other hand, were dissatisfied with the complicated Liverpool system of 27 gradient variations. Even more irritating, an arbitration board located in Liverpool evaluated cotton after its arrival in England, which meant that American interests either had to accept the board's ruling or have their cargo returned to the United States, neither very attractive alternatives in the event of disagreement over grading.[48]

Naturally in favor of gaining adoption of American standards in the international trade, Wallace sent a delegation from the Department of Agriculture to the World Cotton Conference in Liverpool in June 1921 to secure endorsement for such a proposal.

[48] "Universal Standards for American Cotton," ms.; and W. R. Meadows, W. L. Pryor, and Chester Morrill to George Livingston, June 19, 1921, *ibid.*

Despite the active support of private cotton interests from the United States and an interest on the part of groups from a number of other countries, the powerful Liverpool association was again able to prevent action on the matter.[49] After adjournment of the conference, two members of the department's delegation traveled to several European countries attempting to secure acceptance of American standards, seeking thereby to bring pressure on the Liverpool organization. While the cotton importers of Milan, Italy, showed some interest in the American system, other groups chose to follow the lead of Liverpool and rejected suggestions for its adoption.[50]

Little more was done on the question of universal standards until after passage of the Cotton Standards Act in March 1923. Requiring that all American cotton transactions, whether domestic or foreign, be on the basis of grades established by the Agriculture Department, the measure provided for establishment of a board of appeals in the United States to settle disputes over evaluation. Liverpool standards for American cotton, therefore, were in effect prohibited. In Wallace's opinion the act was of "vital importance to the cotton industry," and he used his influence to help bring about its enactment.[51]

The Liverpool Cotton Association's immediate reaction was to

[49] George Livingston to Wallace, April 19, 1921; L. F. Fitts to George Livingston, April 21, 1921; W. R. Meadows, W. L. Pryor, and Chester Morrill to George Livingston, June 19, 1921, Files of the Secretary of Agriculture; and Chester Morrill to Leon M. Estabrook, August 16, 1921, BAE Files.

[50] W. L. Pryor to George Livingston, June 25, 1921; memorandum, Leon M. Estabrook to Wallace, July 16, 1921; Marsilio Volpi to W. R. Meadows and Chester Morrile [sic], June 30, 1921; W. R. Meadows and Chester Morrill to Leon M. Estabrook, undated, Files of the Secretary of Agriculture; and Leon M. Estabrook to Wallace, August 4, 1921, BAE Files.

[51] Wallace to H. M. Lord, December 19, 1922, Files of the Secretary of Agriculture; and Wallace to George W. Norris, February 24, 1923, BAE Files.

threaten to do business only on the basis of its own grading system, which understandably gave American exporters cause for concern. More realistic minds prevailed, however, and the association requested a conference with the Agriculture Department to discuss the question of standards. Wallace consented and set the meeting for June 11 and 12, 1923, in Washington. In the meantime, the BAE held a series of conferences throughout the South to explain its recently revised cotton standards and to rally the support of private interests for the approaching meeting with the Liverpool association. Represented at these sessions were planters, merchants, spinners, and exchanges, the large majority of which agreed to back the Agriculture Department in the negotiations on American standards.[52]

Besides members of the Liverpool group and the Agriculture Department, representatives of the Manchester (England) and Le Havre (France) Cotton Associations and of private interests in the United States were also in attendance at the Washington conference. Soon after the opening of the meeting, a bitter debate broke out in which the European associations objected to the adoption of the American system for the international cotton trade. Complaining that they should have been consulted before the passage of the Cotton Standards Act, Liverpool representatives again threatened to operate only on the basis of their own standards and were supported by the other associations. Although they offered some minor concessions, members of the Agriculture Department, led by Taylor and supported by the entire American delegation, held firm to their position. After a lengthy discussion, the conference, with the Liverpool representatives abstaining, finally decided to accept the American system with some modification of grades. The Liverpool men agreed, however, to submit the conference resolution to their association,

[52] "Report of Special Committee on Revision of the United States Official Cotton Standards," ms. in Files of the Secretary of Agriculture.

which a month later yielded and adopted American standards as the basis for its cotton negotiations. Meeting in Washington on July 17, 1923, a second conference made minor revisions in the grades and prepared for the change-over. On August 1, 1923, the day the Cotton Standards Act went into effect, American standards replaced those of Liverpool in the world cotton trade.[53]

Passage of the Cotton Standards Act and the work of the Department of Agriculture in securing international acceptance of American standards contributed to the stabilization and rationalization of the cotton trade. Planters and merchants could now export their products knowing beforehand the grade and the price it would bring on the foreign market. With the institution of uniform grades for immediate delivery as well as futures transactions, conditions in the domestic trade likewise improved. It was no longer possible for speculators to take advantage of a system of multiple standards throughout the country's exchanges to victimize ill-informed producers.

The Department of Agriculture's efforts to establish and enforce national wheat standards, on the other hand, met with considerable domestic opposition. The Grain Standards Act of 1916 required the use of a uniform grading system established by the department for all grain shipped in interstate commerce. During World War I wheat was handled by the United States Grain Corporation and no attempt was made to enforce standardization. Shortly after the war, the Department of Agriculture formulated a set of uniform standards and endeavored to apply it throughout the country, resistance to which was developing at the time Wallace took office.[54]

The states of Washington, Oregon, and Minnesota were the

[53] Taylor to Wallace, July 26, 1923, *ibid.*; *New York Times,* September 19, 1923, p. 2; and Taylor, "A Farm Economist in Washington," pp. 169–78.

[54] Taylor, *The Story of Agricultural Economics,* pp. 60–66.

centers of strongest opposition. Under the leadership of their State Department of Agriculture, Washington farmers campaigned vigorously against the establishment of uniform wheat standards. Charging that the government supervisors licensed to evaluate grain were too strict and that federal grades were based on an inequitable scale, they insisted on the retention of the state system under which Washington had operated for some time. Joining in the fight against adoption of federal standards was the Portland Chamber of Commerce, which had also established a grading system for grain shipped from its port. A similar situation existed in Minnesota, where farmers favored marketing their wheat under state standards because they were more lenient than those of the Agriculture Department. Likewise opposed to government interference were the major grain exchanges, which preferred to handle their own grading and inspection.[55]

In April and May of 1921 representatives of the Department of Agriculture held several conferences with delegations from the country's wheat regions to discuss possible changes in federal grades. At these meetings farmers demanded either abandonment or a general lowering of standards, while the department took the position that stricter grading would be more beneficial to producers in the long run. Pointing out what he felt was the fallacy in their reasoning, Wallace explained to farmers that "calling a low-grade wheat by the name of a higher grade does not make it a higher grade and will not get a higher price for it. The result will only be to bring down the average price of all the wheat in that grade and penalize the seller of the better wheat." The hearings failed to convince Wallace that changes in standards were in order, and on May 17, 1921, he announced that the prevailing federal grades would be maintained. As a concession to opponents of the department's standards, however, he

[55] Taylor, "A Farm Economist in Washington," pp. 142–50.

promised a thorough study of the question and future modifications if deemed advisable.[56]

As the controversy continued, criticism of the Agriculture Department mounted. In July 1921 a congressman from Washington advised Wallace that "the wheat growers of my section are much disturbed and they sincerely believe that they are discriminated against by the federal grain grades." [57] Representative Halvor Steenerson of Minnesota complained to Harding and Hoover that the regulations of the Department of Agriculture favored the grain trade at the expense of producers.[58] Steenerson introduced a bill into the House which, had it been enacted, would have prescribed wheat grades by legislation and taken the matter out of the hands of the department. Asserting that the formulation and enforcement of standards was "a proper function of executive action" and not a problem to be handled through "rigid legislation," Wallace was unalterably opposed to the measure.[59]

Making good his promise to investigate the question further, Wallace appointed a study committee in late 1921. After examining the case presented by the wheat producers and the views of the department, the committee submitted its findings early the following year. The report recommended several modifications in federal grades and in the inspection system, most of which Wal-

[56] Memorandum, George Livingston to Wallace, April 30, 1921, BAE Files; USDA press release, May 17, 1921, copy in Files of the Secretary of Agriculture; "Report of the Secretary," USDA *Yearbook of Agriculture*, 1922, p. 21; and *Minnesota Farm Review*, January 17, 1924.

[57] John W. Summers to Wallace, July 2, 1921, Files of the Secretary of Agriculture.

[58] Halvor Steenerson to Harding, May 27, 1922, Warren G. Harding Papers, State Historical Society of Ohio, Columbus, Box 2; Halvor Steenerson to Hoover, May 27, 1922, Herbert C. Hoover Papers, Herbert Hoover Presidential Library, West Branch, Iowa, 1-I/258.

[59] Wallace to W. B. Pollock, May 29, 1922, Files of the Secretary of Agriculture.

lace accepted and put into effect in April 1922.[60] Although the concessions failed to meet all of the objections, they were apparently sufficient to placate most wheat producers, for criticism of federal standards were less frequent after the changes had been adopted.

Wallace believed that a rigid federal system of wheat grading was in the best interest of both producers and the grain trade. It would give farmers an accurate idea of the actual worth of their own wheat and provide them with an index through which to determine the price it should bring on the market. Assured of accurate evaluation, buyers could purchase grain with greater confidence, a factor particularly important in stimulating the export trade. The secretary also pointed out that a uniform grading system would expedite the movement of wheat throughout the country and avoid the confusion of different standards for every state or exchange.[61] Despite the resistance of producers, there was small doubt that the department's work along this line benefited the wheat industry.

Under Wallace the Agriculture Department also expanded its inspection and standardization service for perishable goods. When he assumed office, a system for inspecting fruits and vegetables was already in operation, under which department officials at central markets issued certificates attesting to the condition of products upon arrival. Helpful in protecting farmers from false reports by commission men who handled their merchandise at buying points, the system was useless in assessing blame for damage during shipment. The BAE solved this problem by establishing machinery for shipping-point inspection in cooperation with the states. Under the new system certificates of condition were issued before the goods were dispatched, which made it possible to

[60] USDA press release, April 17, 1922, copy in *ibid.*
[61] Speech delivered on January 11, 1924, copy in the Taylor Papers.

prove the liability of shipping agents if damage occurred in transit.[62]

Shipping-point inspection provided service beyond the immediate protection of producers. In issuing certificates and through the elimination of inferior or damaged merchandise prior to shipment, inspectors were able to apply a kind of quasi-standardization to perishables. Although Wallace would have preferred compulsory standards similar to those on cotton and wheat, Congress failed to pass the necessary legislation, and he had to be content with the limited grading possible under the inspection service.[63] Once buyers found that they could depend upon the validity of federal inspection reports, they began purchasing fruits and vegetables through f.o.b. auctions on the basis of the certificates of condition, thus permitting sale at shipping points and direct delivery to retail outlets. As a result, marketing and distribution were greatly facilitated, the possibility of spoilage was reduced, and producers received payment upon shipment rather than after transactions at central markets.[64]

On the basis of its research on the standardization of other products, such as wool, tobacco, and meats, the BAE worked out grading systems and urged their adoption by cooperative marketing associations. Wallace felt that through the use of voluntary standards these organizations could upgrade their own operations, improve the quality of farm products, and benefit the agricultural industry in general. As was its function in other relations with cooperatives, the BAE only supplied the necessary informa-

[62] "Report of the Secretary," USDA *Yearbook of Agriculture,* 1924, pp. 34–35; and Taylor, "A Farm Economist in Washington," pp. 211–12.

[63] Wallace to George B. Christian, Jr., September 20, 1922, Harding Papers, Box 197.

[64] "Report of the Secretary," USDA *Yearbook of Agriculture,* 1924, p. 35; Taylor, "A Farm Economist in Washington," p. 213; and USDA *Annual Report,* 1923, pp. 167–69.

tion and guidance, and left the operational work to private groups. While many farm associations took the initiative and instituted grading systems, the movement was not as widespread as Wallace would have wished.[65]

Another function of BAE was the administration of agricultural extension. Originally under the States Relations Service, responsibility for this work was turned over to the BAE after its creation in 1922. Serving as an information dispensing agency, agricultural extension became an important part of the department's program of farm economics. County agents, through whom the extension service operated, kept the producers in their areas abreast of the latest findings and work of the BAE, instructed farmers in the use of materials issued by the bureau, and encouraged the adoption of practices recommended by the department. The extension service was, in short, the major promotional office for the economic program which the BAE was endeavoring to institute.

While Wallace was secretary, the agricultural extension service came under rather sharp attack. In most states county agents had formed bureaus of local farmers to aid them with their instructional work. This system operated satisfactorily until the postwar period when local groups began organizing into state units and eventually into the American Farm Bureau Federation. Accompanying the federation movement was a change in the program of county bureaus from an educational to a political emphasis. With this change, the relationship between county agents and the bureaus drew the criticism of rival farm organizations and other groups.

In many cases the complaints against the extension service were entirely justified. A county agent in West Virginia, for example, became the chief recruiting officer for the Farm Bureau

[65] "Report of the Secretary," USDA *Yearbook of Agriculture,* 1922, p. 20.

in his area. He set up his office as the headquarters of the local organization and used extension service stationery and franking privileges to solicit members through the mails.[66] In view of the importance of farm bureaus in the work of the county agents, the Missouri state extension director readily admitted, "we cannot hope for such employees to be disinterested parties." Another director in Michigan pointed out that in addition to their educational services, the local bureaus also contributed part of their membership fees to the extension program. "I do not see," he maintained, "how we can do otherwise than give these county membership campaigns our full moral support." [67] Although some state extension divisions, notably in Wisconsin and Pennsylvania, refused to back the federation movement, the rapid rise of the Farm Bureau was due in large part to the organizational work of the county agents at the local level.[68]

Much of the criticism of the Agriculture Department came from the Farmers' Union, whose position was of course threatened by the rise of the Farm Bureau. The president of the Arkansas Union complained to Wallace that "the game as now being played by the Federal employees is an injustice to all self help farm organizations," and he called for "a complete divorce-

[66] R. J. Briant to Farm Bureau Members, Old and New, August 23, 1922; and C. W. Pugsley to N. T. Frame, September 5, 1922, Files of the Secretary of Agriculture.

[67] A. J. Meyer to C. W. Pugsley, September 8, 1922; and R. J. Baldwin to C. W. Pugsley, September 11, 1922, *ibid.*

[68] The extension director of Pennnsylvania formed farmers' advisory committees to replace the county bureaus after the federation movement began. In Wisconsin the agricultural college of the University, which directed the extension work, recognized the pitfalls in basing the service on private groups, and county agents were directed not to form local units. As a result, the Farm Bureau had to do its own organizational work in that state. Gladys Baker, *The County Agent* (Chicago: University of Chicago Press, 1939), p. 20; H. L. Russell to H. G. Smith, November 5, 1918; H. L. Russell to E. A. Girge, February 11, 1920; H. L. Russell to J. R. Howard, April 1, 1920; and H. L. Russell to A. J. Glover, April 12, 1920, Files of the College of Agriculture, Series 9/1/2, Box 15,

ment of the Department and the Bureau Federation." [69] Other farm organizations as well as a number of business groups joined in the attack. There was no argument against purely educational units, they explained, but now that the farm bureaus had become primarily political pressure groups, it was imperative that the county agents sever connections with them.[70]

Concerned over criticism of its extension operations, the department endeavored to remedy the situation soon after Wallace became secretary. In April 1921 Alfred C. True of the States Relations Service met with the executive committee of the American Farm Bureau Federation to draft a "Memorandum of Understanding." Signed by True and federation president James R. Howard, the statement declared that county agents would "aid the farming people in a broad way with reference to problems of production, marketing, and formation of farm bureaus and other cooperative organizations," but that they were to be prohibited from organizing bureaus, soliciting membership, or taking part in "activities which are outside their duties as extension agents." [71] Despite the announced position of the department, many county agents persisted in engaging in farm bureau work and the controversy continued. With little improvement in the situation, Wallace restated in August 1922 that extension employees were to undertake only educational work, emphasizing the prohibition on organizational functions.[72] In the meantime, the matter had at-

[69] A. C. Davis to Wallace, August 24, 1922, Files of the Secretary of Agriculture.

[70] Baker, *The County Agent*, p. 19; and James H. Shideler, *Farm Crisis, 1919–1923* (Berkeley and Los Angeles: University of California Press, 1957), p. 126.

[71] "Memorandum of Understanding," copy in files of the Secretary of Agriculture and reprinted in Alfred C. True, *A History of Agricultural Extension Work in the United States, 1785–1923,* USDA Miscellaneous Publication No. 15, 1928, p. 169.

[72] "Statement of the Secretary of Agriculture Concerning the Relationship of Federal Cooperative Extension Employees to Agricultural Organizations," USDA press release, August 25, 1922, copy in Files of the

tracted the attention of Congress and led to an investigation by the House Committee on Banking and Currency. Disclosing little that was new, the hearings contributed nothing to finding a solution to the problem and seemed to confirm the charges of extension critics. Particularly embarrassing to the Agriculture Department was a statement by Farm Bureau president Howard that the "county agent is the keystone of the federation." [73] Interest in the investigation was never strong, however, and the committee issued no report or recommendations beyond the record of the hearings.

The Agriculture Department's position in the extension service controversy was a difficult one. Although the county agents were its official representatives, the BAE had little control over them. Maintained through a federal-state cooperative arrangement, the extension service was administered at the state end by the agricultural colleges. They appointed the county agents, paid their salaries, and directed their activities in the field, while the function of the BAE was merely to supply educational material and to contribute financially to the extension program. The department, therefore, could issue dicta, but had small means for enforcing them. If the agricultural colleges were unwilling to abide by the policy, there was little the BAE could do to compel compliance.

The conflict over the extension service hampered the economic program of the BAE. As the Farm Bureau came to represent the larger, wealthier producers, the older organizations charged that the Agriculture Department was aiding this class of farmers rather than those most in need of assistance. Viewed by many as adapted primarily to the operations of big producers, extension

Secretary of Agriculture and reprinted in True, *A History of Agricultural Extension Work in the United States, 1785–1923*, p. 170.

[73] United States Congress, House, *Hearings on Farm Organizations*, 67th Cong., 1st Sess., p. 88.

service instruction in farm economics was received with little enthusiasm by a large segment of the farming community. Disagreement over the functions of county agents, furthermore, caused dissension between agricultural colleges and the department and prevented the close federal-state cooperation necessary to insure a successful program.

Unsatisfactory as it was, agricultural extension was the principal and, outside of department publications, the only vehicle for carrying the results of research and study in agricultural economics to farmers in the field. It represented the main link between the BAE and individual producers and, in that capacity, performed an important function. But economic education proved to be a more difficult task than realized, and the extension service was at best an imperfect instrument for fashioning the kind of fundamental modification within the farming industry envisioned by the Agriculture Department.

For his program of agricultural economics, Wallace drew freely upon earlier reform tradition. Just as prewar progressives sought to accommodate American political, economic, and social institutions to advancing industrialism, so he attempted to bring about the adjustment of agriculture to conditions in the twentieth century. Although Wallace was hardly a progressive in the narrow sense of a social reformer, his ideas and methods as applied to agricultural economics reflected broadly the principal elements of reform thought.

As was true of many prewar progressives, Wallace's prejudice against business interests provided the basis for his thinking. An agrarian resentment of the power of industrial and financial groups colored his analysis of American development and shaped his approach to the country's problems. The cause of economic ills, the secretary contended, could be traced to the activities of business interests. Their unwise and abusive use of power had resulted in an unbalanced, disjointed economy in which weaker members were increasingly exploited. If not remedied, this condition would surely lead to serious economic dislocations and, perhaps, to the undermining of American institutions themselves. Like his analysis, Wallace's solution was firmly within the progressive tradition. He recommended that the government take steps to clear away the obstacles to the rise of depressed groups and to equip them to compete successfully within the capitalistic system. Such action, he theorized, would bring about the disappearance, or at least minimization, of the advantage of business-financial interests.[74]

One of the identifying features of the prewar reform tradition was a healthy respect for the social sciences, and earlier progressives were enthusiastic about the social and economic improvement which they felt would accrue from the application of new academic disciplines. Wallace likewise envisioned vast possibilities for agricultural economics. At a time when many colleges were dubious about the practical value of the young science, he was confident that this discipline held the key to the solution of farm problems. Use of the knowledge gained from the study of agricultural economics, moreover, would be a large step toward both the elimination of agrarian inequity and the realization of the social equilibrium which had been a principal objective of the progressive era.

The BAE itself found precedents in the earlier progressive period. Under the governorship of Robert M. La Follette in the first years of the century, a group of University of Wisconsin professors served as advisors to the state government, foreshadowing later brain trusts. Theodore Roosevelt's Bureau of Corporations and Woodrow Wilson's Federal Trade Commission were, in addition to regulatory agencies, organizations of experts for the collection and analysis of data. Created in 1916, the Tariff Commission represented an attempt to remove the tariff issue from

[74] Wallace, *Our Debt and Duty to the Farmer*, pp. 164–65.

politics and deal with it in a more systematic way. The theory behind the establishment of these agencies was that, with more information and greater understanding of the functioning of the American economy, the problems of industrial society could be solved. In creating the BAE Wallace operated from the same premise. He and Taylor assembled a group of outstanding specialists in the various areas of agricultural economics to gather facts and render interpretations. Convinced that well-trained scientists were a national asset, they established a program of inservice training to prepare new recruits in the approach of the department and to keep veteran members abreast of the latest developments in the field.[75] Representing an important element of reform thought both before and after the war, faith in expertise became central to Wallace's program of agricultural economics.

Along the same line, Secretary Wallace had a progressive interest in administrative efficiency and did much to improve the operations of his department. Besides the BAE, he established the Bureaus of Home Economics and Dairying for the purpose of handling the department's work along these lines more effectively. Among the various bureaus, and between them and the secretary's office, lines of authority were clearly drawn and responsibility precisely delegated. Although the bureau chiefs had considerable latitude and control over their own divisions, Wallace demanded frequent and rather extensive reports on their work. To acquaint the bureaus with the projects of each other and to coordinate their activities, the secretary also held weekly meetings of the chiefs.[76]

Throughout the rhetoric of farm representatives during the 1920's ran evidence of concern over the loss of agrarian status.

[75] E. D. Ball to J. H. James, April 24, 1922; E. D. Ball to Don B. Colton, September 13, 1921, Files of Secretary of Agriculture.

[76] Memorandum, W. A. Jump to Bureau Chiefs, September 13, 1921, *ibid.*

Though not as well articulated or even consciously presented as complaints about the economic conditions of farmers, anxiety to regain the position of influence and prestige once enjoyed by the yeoman class was nevertheless a crucial factor in rural thinking. As an aggressive representative of agricultural interests, Wallace was particularly sensitive to the relative decline of the agrarian class within the country's social structure. He lamented that farmers were no longer consulted, that little attention was paid to their views. Rather than playing an important role in the determination of national policy as they had at one time, farmers now exercised small influence in the process of decision-making. Just as many progressives of the reform era had designed a movement to restore them to a lost position in society, so Wallace offered his program not only as a solution to the farmers' economic ills but as a way of reviving the once proud yeoman heritage.[77]

And, finally, progressivism had one further identifying feature which was also common to Wallace's thinking—a decidedly moral strain. Rural civilization, according to the secretary, had produced a special breed of men who were strong, vigorous, and self-reliant. Removed from the mob psychosis of the city, farmers were more stable and less likely to be captured by misguided popular movements than their urban counterparts. Relative isolation from the corrupting influence of the city and the soul-enriching experience of making a living from the soil produced in farmers a temperament and spirit which made them the most dependable element in society. For this reason, Wallace asserted, the agricultural class was the balance-wheel of American civilization, the keeper of the country's values, and the force on which the government depended in time of crisis. Could the country well afford to stand idly by and watch the dissolution of the bul-

[77] *Wallaces' Farmer* 44 (August 29, 1919): 1652; and address before the Boston Chamber of Commerce, December 19, 1921, copy in Taylor Papers. Concern over loss of agrarian status was a major theme of Wallace's book, *Our Debt and Duty to the Farmer*.

wark of American democracy? Wallace thought not. And his form of progressivism suggested to him a way to save the farming class not solely for its own sake but for the good of the country in general.

In spite of Wallace's high expectations, agricultural economics failed to rejuvenate farmers either economically or socially. Like many reforms of the progressive era, his program proved inadequate to meet the problems of advancing industrialism. Wallace placed considerable faith in the ability and willingness of farmers to use the material and instruction provided for them to control their own destiny. Assuming that supplying them with economic information was equivalent to giving them an economic education, he missed the fact that farmers had neither the organization, machinery, or training necessary to put the research and statistics of the BAE to really effective use. Wallace failed to understand that the nature of agriculture, based as it was on individual enterprise, prevented it from successfully utilizing the methods of big business. Even with the help of the BAE, farmers were unable to exercise the same control over their production, marketing, and distribution as industrial interests. The secretary, furthermore, lacked an appreciation of the plight of the poor farmers and tenants who, no matter how much inclined they might have been to accept the department's program, found it impossible to practice such things as production control and orderly marketing.

Although most farmers endorsed the movement for agricultural economics when Wallace first took office, disillusionment soon set in. Without the anticipated improvement in conditions, producers were drawn increasingly toward more radical solutions. Though he never abandoned his program, Wallace himself came to the conclusion that more was needed if American agriculture were to recover its proper place in the economy.

Chapter VII / THE NATIONAL AGRICULTURAL CONFERENCE OF 1922

Wallace believed that farmers were not only capable of making intelligent and effective use of the services of the BAE, but that they could also formulate realistic solutions to their problems. The difficulty lay in finding a vehicle for expressing the wishes of producers and exposing those in positions of authority to their ideas. As editor, Wallace had maintained that farm representatives in Congress should not claim to be spokesmen for agrarian sentiment in every instance. "The moment they do this," he cautioned, "they are influenced by their own personal opinions, which in turn are modified by their personal contacts in Washington." Rather, they should periodically consult producers to learn their attitudes and thinking on matters affecting agriculture. "When it comes to voicing farm sentiment," Wallace thought, ". . . that can be done very much more effectively by calling practical farmers and farm representatives in from the country." [1]

This is exactly what Wallace had in mind when he called the National Agricultural Conference of 1922. During the campaign of 1920 he had recommended to Harding that if elected he should summon a conference to discuss the farm situation and

[1] *Wallaces' Farmer* 45 (May 21, 1920): 1419.

possible solutions to it. Throughout 1921 Wallace renewed his suggestion on several occasions, but found the President unreceptive. Secretaries Herbert Hoover and Andrew Mellon insisted that a farm conference was unnecessary, and Harding, himself unable to see any value in a meeting at the time, chose to follow their advice on the matter. Apparently the success of the farm bloc and the rather pessimistic report of the Joint Commission of Agricultural Inquiry in October 1921 influenced the President's thinking, for, after a game of golf in late December 1921, Harding casually told Wallace to proceed with plans for an agricultural conference.[2]

A few days later the President sent Wallace an official request to call the conference, with suggestions on how he should proceed. Feeling that the meeting might "be a very helpful agency in suggesting practical ways of improvement," Harding recommended that the conference coordinate its work with the investigation recently completed by the Joint Commission of Agricultural Inquiry. In addition to the "ablest representatives of agricultural production," he directed Wallace also to invite "those who are engaged in industry most intimately associated with agriculture." Attendance of interests "closely related and dependent upon agriculture," the President thought, "will clarify our views. . . ." The final make-up of the conference and the selection of delegates, however, he left to Wallace's discretion.[3]

Selection of delegates to the conference proved to be a difficult and rather trying task. While Wallace made every effort to strike a balance among the groups invited to send representatives, he

[2] Henry C. Taylor, "A Farm Economist in Washington," p. 242, ms. in Henry C. Taylor Papers, State Historical Society of Wisconsin, Madison.

[3] Harding to Wallace, December 30, 1921, Warren G. Harding Papers, State Historical Society of Ohio, Columbus, Box 197 and Files of the Secretary of Agriculture, National Archives; also in United States Congress, House, *Report of the National Agricultural Conference, 1922,* Document 195, 67th Cong., 2nd Sess., pp. 3–4.

found it impossible to satisfy all factions. It would have been difficult enough to placate the various farm organizations and agricultural interests, but Harding's order that spokesmen for business should also be included complicated the matter immensely. Wallace received a large amount of correspondence recommending, requesting, and demanding that certain groups and interests be permitted to send delegations. In order to meet the wishes of everyone, the secretary would have had to invite so large a number as to make the conference unmanageable. Interested in satisfying as many elements as possible, however, Wallace conferred with farm organization leaders, agricultural colleges, and advisors in his department before making a final decision on the make-up of the meeting.[4]

As might well be expected, there was a good deal of criticism of the conference even before it opened. The Farmers' National Council, a lobbying agency maintained in Washington by several of the more militant agricultural organizations, decided that nothing could be accomplished if interests which exploited farmers were invited.[5] "Your great men," wrote a disgruntled correspondent, "will get together and make fine speeches of which there will be self and mutual admiration but we farmers will have to drag it out the same as before. . . ." [6] Many were convinced from the outset that the conference was merely an administration device to divert criticism of its agricultural policy and to rescue leadership from the farm bloc. Amid numerous charges that the meeting would be "hand picked," one critic wondered whether arrangements were being made "with a purpose to cut and dry the whole matter, to severely limit the list and to ignore organizations which are not regarded as 'conservative,'" while another feared that real "dirt farmers" would be inadequately

[4] *Washington Post,* January 26, 1922, p. 1.
[5] *New York Times,* January 1, 1922, p. 14.
[6] John Tegley to Wallace, January 6, 1922, Files of the Secretary of Agriculture.

represented.[7] Sensitive to the barrage of complaints, Wallace carefully responded to all inquiries on the selection of delegates. "In making up the personnel of the conference," he explained in a typical reply, "we have done all that is humanly possible to insure that every phase of the agricultural interests of the Nation is adequately represented." [8] Criticism persisted, however, and the conference itself only served to strengthen the opinion of some that the delegates had indeed been specially selected.

Actually, the agricultural conference was probably as well-balanced as possible. A sizable bloc of the 336 delegates was composed of representatives from some 20 different farm organizations, including the militant Farmers' Union, the Grange, and the Farm Bureau. Significantly, individual farmers not officially representing any specific organization, but chosen on the recommendations of farm groups and agricultural colleges, made up the largest faction. Such specialized groups as farm women and Negro farmers were also given a voice at the conference, as well as officials of state agricultural departments, farm editors, and members of agricultural colleges. Also in attendance were delegates sent by food distributors and processors, implement manufacturers, and a single representative of organized labor.[9] However unsatisfactory the effort might have seemed to some, Wallace made an attempt to be impartial in the selection of those invited to attend the agricultural conference.

At the time Harding gave his consent for the conference, Taylor was attending a meeting of economists in Philadelphia. In a

[7] A. D. Fairbairn to Wallace, January 2, 1922; Fred McCulloch to Wallace, January 18, 1922, Files of the Bureau of Agricultural Economics, National Archives.

[8] Wallace to Noel Ray, January 12, 1922, *ibid.*

[9] Note in *ibid.* The breakdown of representatives to the conference was: individual farmers, 108; representatives of farm organizations, 79; farm women, 10; Negro farmers, 5; state agricultural officials, 15; farm editors, 22; members of agricultural colleges, 30; representatives of business groups, 62; representatives of labor, 1; others, 4.

wire to the bureau chief, Wallace instructed him to obtain suggestions from those at the meeting on possible topics to be included in the conference program. Taylor consulted Richard T. Ely and John R. Commons of the University of Wisconsin, George F. Warren of Cornell, Wesley C. Mitchell of the New York Bureau of Economic Research, and a number of other prominent economists, all of whom had not only academic interest in the agricultural crisis but also deep concern over the rural situation. Most felt that if the conference were to produce any worthwhile results it should concentrate on matters dealing with farm credit, cooperation, and agricultural economics. Consistent with the ideas of Wallace and Taylor, these recommendations were acceptable and the Agriculture Department generally followed them in making up the agenda for the conference.[10]

Wallace realized that if the agricultural conference were to produce any tangible benefit to farmers, its proposals would have to be practical and at the same time acceptable to the administration. As he and Taylor worked out the details, therefore, the secretary kept in close contact with Harding. Of the opinion that the conference might well "bring forth the general outline of a national program which may assume a very considerable authority and importance hereafter," the President urged that care be taken "to insure that such a program would be based on entirely sound and safe constitutional principles." [11] Clearly concerned lest militant elements gain control of the conference, Harding was anxious that there be established safeguards to prevent the adoption of recommendations which might prove embarrassing to the administration. Thus Wallace and Taylor chose to control the meeting, not through hand-picked delegates, but by drafting a program and set of procedural rules which offered little oppor-

[10] Taylor, "A Farm Economist in Washington," pp. 243–44.
[11] Harding to Wallace, January 11, 1922, Henry C. Wallace Papers, University of Iowa Library, Iowa City, Box 3.

tunity for the passage of unwelcome resolutions. Under their plan the major work of the conference was to be carried on in prearranged committees, each assigned to a particular aspect of the farm problem, and the speeches from the main floor were to be limited to five minutes.[12]

In working out the final program, Wallace encountered a bit of a problem with his oldest son. Invited as a representatives of the Corn Belt Meat Producers' Association, Henry Agard and A. Sykes were both scheduled to address the conference. With little confidence in Sykes' ability to work up a satisfactory speech, Wallace suggested to his son that he help him with it. Accepting the task with more enthusiasm than the father had anticipated, Henry Agard wrote a paper the major part of which was devoted to an attack on the permanent tariff bill then under consideration by the Senate Finance Committee. This would not do, of course, and Wallace had to persuade his son and Sykes to delete the section on the tariff from the speech. Prior to leaving for Washington, however, Henry Agard had decided to print the entire speech in *Wallaces' Farmer,* and the issue was on its way to press before his father saw the draft. As a result, the attack on the administration's tariff policy appeared in the journal soon after the close of the conference. Despite the fact that Wallace had disavowed any responsibility for the editorial policy of the paper when he took office, Harding was nevertheless disturbed over what appeared to be criticism from within his cabinet. Wallace's only comment to his son was "Henry, have a heart." [13]

The National Agricultural Conference convened on January

[12] Wallace to Harding, January 11, 1922; Wallace to H. A. Wallace, January 9, 1922, BAE Files; and interview with Donald R. Murphy, Des Moines, Iowa, August 29, 1965.

[13] Wallace to H. A. Wallace, January 9, 1922, BAE Files; *Wallaces' Farmer* 47 (February 3, 1922): 135; and Henry A. Wallace, Review of James H. Shideler, *Farm Crisis, 1919–1923* in the *American Historical Review* 63 (April, 1958): 707.

23, 1922, for a five-day meeting. At Wallace's request Representative Sydney Anderson, head of the Joint Commission of Agricultural Inquiry, acted as permanent chairman. Divided into 12 committees, the conference devoted the mornings to speeches by various representatives to the combined delegation and the afternoons and evenings to the small-group meetings. At the sessions of the individual committees, members discussed their assigned topics and drafted appropriate resolutions, upon which the whole conference voted in the final session.

President Harding gave the opening address and set the mood for the conference. Calling for "the proposal of feasible and practicable methods for dealing with . . . [agricultural] conditions," he suggested that these should include recommendations for improved rural credit facilities, aid to cooperatives, and expansion of the federal farm economics program. "It can not be too strongly urged," he hastened to add, "that the farmer must be ready to help himself." [14] The President sought to impress upon the conference that, while government assistance might prove useful, "legislation can do little more than give the farmer the chance to organize and help himself." But the part of Harding's speech that provoked the most interest and resentment was his passing reference to the farm bloc. Speaking of the poor state of agriculture, he expressed the idea that it was "truly a national interest, and not entitled to be regarded as primarily the concern of either a class or section—or bloc." Reference to the Senate group was apparently an inspiration of the moment, for it was not originally part of the speech nor did it appear in the printed text included in the conference report.[15]

After the President's speech, Wallace gave a short address of welcome to the group and explained the procedures. "It is not my purpose to tell you what you should do here . . . ," the secre-

[14] *Report of the National Agricultural Conference, 1922*, pp. 6–13.
[15] *New York Times,* January 24, 1922, p. 1.

tary told the delegates, but he indicated that, in the interest of expediency, his department had felt it wise to lay the groundwork for the conference. Outlining the agenda, he pointed out that there would be two objectives: first, to deal with the immediate emergency and possible remedies; second, to consider the problem of establishing a permanent federal agricultural policy. "In a gathering of this kind, composed as it is of strong, independent thinkers," Wallace concluded, "progress must be made through committee action." [16]

Reaction of the more militant representatives at the conference to the two opening speeches was anything but favorable. Their suspicion that the conference was summoned merely to endorse an administration program had been confirmed. To them Harding's call for "feasible and practicable methods" for attacking farm problems was obviously designed to discourage proposals which they considered best calculated to solve the agricultural crisis. Particularly irritating was the President's extemporaneous attack on the farm bloc. Critical of Wallace's plan for the introduction of resolutions through committees rather than from the conference floor, dissatisfied elements charged that the secretary had packed the committees in order to prevent action on unwanted recommendations. J. S. Wannamaker, president of the American Cotton Association and one of the most outspoken critics at the conference, proposed the creation of an additional group to handle resolutions offered outside of the regular committees, and representatives of the Farmers' Union led a move to change the rules to permit the introduction of proposals from the floor. Wallace's control of the conference proved to be firm, however, and neither suggestion was acted upon.[17]

[16] *Report of the National Agricultural Conference, 1922*, pp. 13–15.
[17] *New York Times*, January 25, 1922, p. 8 and January 26, 1922, p. 3; and *Washington Post*, January 25, 1922, p. 3. Wallace remembered later that Wannamaker had "made a good deal of trouble . . . [but] contributed nothing of value." Wallace to George B. Christian, Jr., July 21, 1922, Harding Papers, Box 197.

Much of the controversy at the conference revolved around efforts to pass a resolution recommending government price-fixing. Since the Farmers' Union had come to look upon this solution as a panacea for all agricultural ills, its representatives were determined to gain endorsement of a price-fixing measure. Firmly opposed to the idea, Wallace was equally determined to prevent the adoption of such a resolution by the conference. Thus the stage was set for a clash as the meeting opened.

The committee on marketing of farm products was the scene of most of the discussion on price-fixing. Introduced by John A. Simpson, president of the Oklahoma Farmers' Union, one resolution directed the President and Congress to "take such steps as will immediately guarantee a price below which no farmer in this country will have to sell his 1922 crop of corn, wheat, and cotton." [18] Former Secretary of Agriculture Edwin T. Meredith submitted a similar proposal calling for the creation of a government price-fixing agency.[19] Moving quickly to counteract these efforts, Wallace held an impromptu press conference at which he reiterated the administration's stand on price-fixing, and Taylor was able to convince Meredith to substitute a resolution calling for a special price study in place of his original proposal.[20] At this juncture help came from an unexpected source. George N. Peek, president of the Moline Plow Company, submitted an alternative resolution to the committee, which stated that "whereas the prices of agricultural products are far below the cost of production . . . , the Congress and the President . . . should take such steps as will immediately reestablish a fair exchange value

[18] George N. Peek to F. C. Grether, March 1, 1922, George N. Peek Papers, Western Historical Manuscripts Collection, University of Missouri Library, Columbia.

[19] *New York Times,* January 26, 1922, p. 1; and Taylor, "A Farm Economist in Washington," pp. 244–45.

[20] *Des Moines Register,* January 27, 1922, p. 1; *New York Times,* January 27, 1922, p. 3; and Taylor, "A Farm Economist in Washington," p. 245.

for all farm products with that of all other commodities." [21] Satisfied with this innocuous suggestion, Wallace quietly indicated the administration's endorsement of the Peek resolution. As a result, the committee defeated Simpson's price-fixing measure and accepted Peek's proposal, which was eventually adopted by the conference as a whole.

After the marketing committee rejected the price-fixing proposal, Wannamaker attempted to introduce it from the floor of the conference, but he was blocked by the procedural rules Wallace and Taylor had formulated for just such an occasion.[22] With this failure to gain consideration of their pet measure, dissatisfied elements threatened to form a "rump" convention of "real dirt farmers" to adopt resolutions which had been rejected by the regular meeting. Immediately following the close of the final session, disgruntled delegates made good their threat. Calling themselves the "agricultural bloc" of the conference, they met and endorsed several proposals, including not only government price-fixing but the Norris bill, lower freight rates, and government operation of the railroads.[23]

George N. Peek had a special interest in the conference. He and Hugh S. Johnson, general counsel for the Moline Plow Company, had formulated a plan to improve the purchasing power of farm commodities through the creation of a government export corporation to dispose of agricultural surpluses abroad.[24] Since the rather complicated scheme was not well known nor understood, the two authors decided that the conference offered an excellent opportunity to explain the export corporation idea and

[21] *Report of the National Agricultural Conference, 1922*, p. 171; and *New York Times*, January 27, 1922, p. 1.

[22] *New York Times*, January 27, 1922, p. 1; and *Washington Post*, January 27, 1922, p. 1.

[23] *New York Times*, January 26, 1922, p. 1; and *Washington Post*, January 28, 1922, p. 4.

[24] *Equality for Agriculture* (Moline: H. W. Harrington, 1922). The plan will be dealt with fully in the chapter on the McNary-Haugen bill.

gain publicity for it. Peek requested an opportunity to present the plan to Wallace before the meeting, but the secretary failed to respond.[25] Shortly before the opening of the conference, Peek and Johnson called on Wallace to obtain his reaction to a brief of their proposal they had sent to him earlier. Much to their chagrin, he had not examined the scheme and refused to consent to its presentation at the meeting.[26] In a final attempt to gain permission to place their plan before the conference, Peek and Johnson approached Sydney Anderson. "He frankly told us," Johnson remembered later, "that if we tried to raise this point in that convention, he would steam roller us—all of which he faithfully and artistically did." [27] Unable to introduce the export corporation scheme, the two men had to settle for endorsement of Peek's price resolution, which merely called attention to the condition their plan was designed to remedy. The episode would not have been significant, except for the fact that Wallace was eventually to turn to the "Peek plan" after other solutions to the farm problem had failed.

Samuel Gompers, president of the American Federation of Labor and the single representative of organized labor, also provided the conference with considerable controversy. Placed before the general meeting by the committee on transportation, one resolution called for railroad wage reductions and repeal of the Adamson law, which provided for an eight-hour workday for railway employees. In an impassioned attack on the proposal which extended beyond the five-minute limit, Gompers cautioned the delegates that if it were passed "we [labor] can't help but

[25] Telegram, Peek to Wallace, January 13, 1922, Peek Papers. Supporters of the export corporation plan urged that it be placed before the conference. See F. R. Todd [?] to Wallace, January 3, 1922; and W. M. Ritter to Peek, December 31, 1921, *ibid.*

[26] Gilbert C. Fite, *George N. Peek and the Fight for Parity* (Norman: University of Oklahoma Press, 1954), pp. 44–48.

[27] Hugh S. Johnson, *The Blue Eagle From Egg to Earth* (Garden City: Doubleday, Doran, 1935), p. 106.

regard you as enemies to the working classes of the country." [28] Although he succeeded in having the measure reported back to the committee for rewriting, the revised version was hardly more satisfactory to the labor leader. "We insist," the resolution concluded, "that the railroad corporations and railroad labor should share in the deflation in charges now affecting all industries." Adopted over the loud objections of Gompers, the resolution was bitterly resented by labor.[29]

The agricultural conference passed 37 resolutions, all of which were entirely acceptable to the administration. Recommendations to Congress included proposals for intermediate credit legislation, antitrust exemptions for cooperatives, tariff revision on farm products, stimulation of the agricultural export trade, and the investigation of plans for monetary stabilization. Urging the Agriculture Department to expand and refine its research and service in the area of agricultural economics, the conference advised farmers, for their part, to cut costs through increased operational efficiency, to undertake diversification, and to adjust production to market conditions.[30] Consistent with Wallace's preference for limited federal assistance and voluntary adjustment, the recommendations were likewise in accord with the President's request to place the emphasis on self-help. Militant elements gained a small victory over Harding with adoption of a resolution commending the members of the farm bloc "who, regardless of party, so early saw the emergency and have so consistently supported a constructive program for the improvement of agriculture and the bettering of rural life." [31] All in all, though, the administration could be, and was, quite satisfied with the results of the agricultural conference.

[28] Quoted in the *New York Times,* January 28, 1922, p. 3.
[29] *Report of the National Agricultural Conference, 1922,* p. 142; and *Fort Dodge Messenger,* January 28, 1922, p. 1.
[30] *Report of the National Agricultural Conference, 1922,* pp. 137–84.
[31] *Ibid.,* p. 138.

Assessment of the conference varied widely. Even before adjournment Benjamin C. Marsh, head of the Farmers' National Council and the instigator of the rump convention, pronounced it "a complete and unqualified failure as far as securing results are concerned." [32] Iowa Farmers' Union president Milo Reno thought that the conference "was a wonderful exhibition of a machine well greased to carry out the desires of those who planned it." [33] Convinced that the whole affair was a deliberate attempt to engage labor and farmers in "mutual recrimination" for the purpose of preventing their alliance against common enemies, Gompers could not "recall a conference at which there were greater restrictions placed upon free expression." [34] Other critics saw the meeting as proof that the Agriculture Department was in league with the "money changers." "That was a good meeting and there was [sic] a lot of good talks," wrote a disillusioned farmer, ". . . but Mr. Wallace the time has come when fine talk will not pay the farmers' debt, nor will [it] support the farmers' families." [35]

There were some, on the other hand, who saw the results of the conference in quite a different light. Richard T. Ely, a prominent economist from the University of Wisconsin, found it "enlightening" and felt that the resolutions which were adopted represented a realistic approach to the problems of farmers.[36] One of the country's leading financiers, Bernard M. Baruch, pronounced the meeting a "decided success," but cautioned that "its success

[32] *New York Times,* January 26, 1922, p. 3.

[33] Unidentified newspaper clipping in the BAE Files.

[34] *American Federationist* 29 (March, 1922): 180; and Samuel Gompers, *Seventy Years of Life and Labor* (New York: E. P. Dutton, 1925), vol. 2, p. 521.

[35] W. S. S. Shearer to Arthur [?], January 29, 1922; Shearer to Rexford L. Holmes, June 17, 1922, BAE Files; and Levi Seass to Wallace, February 1, 1922, Files of the Secretary of Agriculture.

[36] Richard T. Ely, "The National Agricultural Conference," *Review of Reviews* 65 (March, 1922): 271–72.

will not be permanent unless there is placed upon the statute books . . . some further legislation." His only disappointment was that Peek and Johnson had been denied the opportunity to present their plan.[37] Even the troublesome Wannamaker decided that if the conference recommendations were acted upon, it would "undoubtedly be of benefit in the gradual reconstruction and rehabilitation of the American agricultural industry." But he was quick to add that "no recommendation was passed that will bring immediate relief." [38]

Secretary Wallace was well satisfied with the conference and declared it a "decided success." To Edwin T. Meredith he wrote that "it was the voice of organized agriculture, and as such, that voice will be heard throughout the country, as well as here in Washington, and the results should be decidedly helpful in working our way out of this severe agricultural depression." [39] As for the qualifications of the delegates, Wallace declared that "never before in our history was there brought together a group of men who so completely represented the agricultural thought and practice of the Nation." [40] Although it is unlikely that he would have invited representatives of business had Harding not requested it, he nevertheless justified their presence at the conference. In view of the close relation between related industries and agricultural production, he argued, the cooperation of these groups was imperative if farm conditions were to be improved. Besides, the association of diverse interests at the meeting, according to the secretary, had led to a better understanding among them.[41] So pleased was Wallace with the conference that he enthusiastically

[37] Bernard Baruch to Wallace, March 1, 1922, BAE Files.

[38] J. S. Wannamaker to Wallace, February 17, 1922, *ibid.*

[39] Wallace to Meredith, February 9, 1922, Edwin T. Meredith Papers, University of Iowa Library, Iowa City, Box 31.

[40] Wallace to Harding, February 6, 1922, Files of the Secretary of Agriculture; also reprinted in *Report of the National Agricultural Conference, 1922,* pp, 4–5.

[41] Wallace to W. A. Macpherson, February 13, 1922, BAE Files.

publicized its results and accelerated the work of his department in implementing its recommendations.[42]

And well Wallace might have been gratified with the National Agricultural Conference, for he had succeeded in guiding it in the desired direction and containing it within the desired limits. As he had anticipated, the scheme of permitting the introduction of resolutions only through committees proved effective in preventing the adoption of unwanted recommendations. While the delegates were not hand-picked, it soon became clear that the Agriculture Department had control of the key committees, as was skillfully demonstrated in the committee on marketing of farm products. Ready for a fight over price-fixing, Wallace artfully blocked the resolution when it appeared. Charles G. Dawes, Harding's Director of the Budget and soon to become Vice President, was favorably impressed with the way the Agriculture secretary handled the situation. "The tact, patience and ability of Mr. Wallace," he reported to the President, "eased off the situation, which might have caused a wide-spread and acrimonious discussion in the country, all to no purpose." [43] Speaking of the conference many years later, Henry Agard Wallace remembered that "my father and Dr. Henry C. Taylor . . . were 'babying' things along, allowing the farmers to blow off steam. . . ." [44] That the program recommended by the group was with minor exceptions one which conformed closely to his own ideas is probably the best evidence of the secretary's domination at the agricultural meeting. Taylor admitted as much when he wrote that "while he [Wallace] kept in the background he was completely in charge from beginning to the end." [45]

Wallace's satisfaction notwithstanding, the conference could

[42] Wallace to Staff, April 18, 1922; and Wallace to Taylor, April 24, 1922, Wallace Papers, Box 3.
[43] Dawes to Harding, February 20, 1922, BAE Files.
[44] Henry A. Wallace, Review of Shideler, *Farm Crisis,* p. 707.
[45] Henry C. Taylor, "Henry C. Wallace," p. 4, ms. in Taylor Papers.

point to few real accomplishments. True, several of its recommendations were actually carried out, but hardly at the instigation of the conference. For one thing, Congress had earlier taken steps to institute similar proposals in response to the work of the farm bloc and the suggestions of the Joint Commission of Agricultural Inquiry. Although the enactment of the Capper-Volstead Act and the Fordney-McCumber Tariff corresponded to suggestions by the conference, the legislation was pending at the time the meeting convened. The Intermediate Credits Act of the following year owed more to the work of the joint commission and the Agriculture Department than to the conference, and Wallace's office had already set in motion a plan to expand and improve its farm economics program. If the meeting performed any useful service at all, it was in publicizing the dire conditions existing in the agricultural industry. Although the joint commission had previously conducted a much more comprehensive study of the matter, the conference report attracted greater public interest and was thus more effective in calling attention to the plight of farmers. But this was of small value without the proposal of realistic solutions, and in the last analysis Taylor's later assessment was probably more accurate than any made at the time. Writing in 1925, he noted that the National Agricultural Conference "met, conferred and adjourned without making a single recommendation which struck at the heart of the problem, except the recommendation that the problem of reestablishing price ratios be carefully studied." [46]

[46] Taylor, "A Farm Economist in Washington," p. 249.

Chapter VIII / A THREAT TO THE FOREST SERVICE

WALLACE, like many progressives, was an ardent conservationist. Having inherited many of his attitudes on conservation from Uncle Henry, the secretary had long been alert to the problems of resource utilization and planning. As personal friends of Gifford Pinchot, one of the leading conservationists of the day, both Wallaces were very much influenced by his passionate views on the subject. During the celebrated fight in 1909–1910 between Pinchot, then head of the Forest Service, and Secretary of the Interior Richard A. Ballinger over development of government-owned coal reserves in Alaska, the Wallaces actively supported the Chief Forester. Since its founding, moreover, *Wallaces' Farmer* had followed the line of the conservationists in its editorial policy and consistently backed their programs. Although Harding's cabinet was generally unsatisfactory to those interested in regulated and scientific resource development, Wallace was a significant exception. In fact, his support of Pinchot in 1909–1910 and his well-known endorsement of conservationist policies made him not only acceptable but one of their first choices for the post. Little did either Wallace or the conservationists realize just how much his views on resources management would be put to a test in the new administration.[1]

[1] Interview with Donald R. Murphy, Des Moines, Iowa, August 29,

Location of the Forest Service was a matter of deep concern to friends of conservation. In 1891 the first forest reserves had been created by the federal government and placed under the supervision of the Department of the Interior. Oriented toward a preservationist policy, the administration of the Interior Department was unsatisfactory to conservationists, who favored rational management and scientific development of the country's natural resources. Led by devotees like Pinchot and Henry Graves, the conservationists agitated for the transfer of control over the forest reserves to the Agriculture Department, which they felt better represented their views. With the succession of Theodore Roosevelt to the Presidency, they had one of their own in the White House and succeeded in bringing about the transfer in 1905. By act of Congress, the reserves became national forests and were placed under the administration of the Forest Service, which was to be a division of the Agriculture Department. Conservationists were of course elated, but members of the Department of the Interior accepted the transfer reluctantly and thereafter watched for opportunities to regain control of the national forests.[2]

Overlapping areas of administration between the Departments of the Interior and Agriculture tended to aggravate bitterness over the control of the national forests and make relations even more vexatious. For one thing, the Interior had charge of the national parks, which were often partially or entirely within the forests. It also handled the construction and management of irrigation and reclamation projects, but investigation and demonstration of irrigation methods and determination of water rights were the responsibility of Agriculture. The Interior Department administered grazing rights on government range land and

1965; and William B. Greeley, *Forests and Men* (Garden City: Doubleday, 1957), pp. 95–96.

[2] Samuel P. Hays, *Conservation and the Gospel of Efficiency* (Cambridge: Harvard University Press, 1959), pp. 36–44.

had control over land claims on the public domain outside of the forests, within which jurisdiction for both belonged to Agriculture. And, except for the national forests, the Interior was in complete charge of the entire federal estate.[3] Ripe for conflict, the situation broke into a bitter interdepartmental fight in the early 1920's.

Harding's appointment of Albert B. Fall as Secretary of the Interior did little to reassure the conservationists. An expert on laws governing public lands, oil, and mining, Fall would seem to have been a logical choice for the position. As an Arizona senator, furthermore, he had taken an active interest in legislation relating to the Interior Department and utilization of natural resources. On the other hand, Fall had revealed decidedly anti-conservationist views in his statements and voting record in Congress, which understandably alarmed those interested in regulated development of the country's wealth. Expressing the feelings of many of his fellow conservationists, Pinchot quipped that "it would have been possible to pick a worse man for Secretary of the Interior, but not altogether easy." More prophetic than even he realized, the former Chief Forester wired a friend after learning of Fall's appointment: "Trouble ahead."[4]

The new administration had been in office barely two months when Fall confirmed the fears of the conservationists by drawing up for Harding's signature an order transferring the Forest Ser-

[3] Cattle grazers often petitioned the Interior Department to intervene in their behalf with the Forest Service in an effort to gain special grazing privileges in the national forests. See for instance William H. McCullough to Dan M. Jackson, July 27, 1921; Dan M. Jackson to Charles V. Safford, July 29, 1921; Charles V. Safford to Dan M. Jackson, August 2, 1921; and Charles V. Safford to Charles M. Crossman, January 4, 1922, Albert B. Fall Papers, University of New Mexico Library, Albuquerque, Correspondence relating to Agriculture and Forests.

[4] Pinchot to Samuel Lindsay, March 6, 1921; Pinchot to Lawrence Houghteling, February 21, 1921; and telegram, Pinchot to Walter Darlington, February 24, 1921, Gifford Pinchot Papers, Manuscript Division, Library of Congress, Boxes 237, 239.

vice to his department. Little concerned with the question of resource development, the President would have preferred administration of the national forests by the Department of the Interior simply because that division controlled the rest of the public domain.[5] But to accomplish this through executive order against the wishes of conservationists would have invited more criticism than Harding cared to risk, at least at the outset of his administration, and he therefore chose to place the matter in the hands of the Joint Committee on Reorganization. Created in December 1920, this congressional committee was assigned to investigate governmental organization and to make recommendations for improving efficiency. An amendment to the enabling act in May 1921 permitted the President to appoint a representative of the executive branch to work with the committee and serve as its chairman. Harding selected Walter F. Brown, a private citizen from his native state of Ohio, for the position. In leaving the transfer question up to the committee, the President had avoided an immediate decision, but he also opened the issue to a spirited and prolonged public debate.

Failing to accomplish his purpose through executive order, Fall, reportedly with Harding's blessing, turned to Congress.[6] Although his ultimate objective was to gain transfer of the entire Forest Service, he concentrated initially on securing legislation which would give his department control over the Alaskan national forests. Applying an argument used by conservationists in the 1890's against the Interior Department, Fall charged that the Agriculture Department was preventing effective utilization of Alaskan timberland through its needless and unwise restrictions on exploitation. The Interior secretary maintained that the pres-

[5] Harding to Pinchot, November 18, 1922, Warren G. Harding Papers, State Historical Society of Ohio, Columbus, Box 701; and Burl Noggle, *Teapot Dome: Oil and Politics in the 1920's* (Baton Rouge: Louisiana State University Press, 1962), p. 23.

[6] *New York Times*, July 17, 1922, vol. 7, p. 2.

ervationist views of the Forest Service had resulted in a maze of impediments which discouraged private development of the territory's national forests. Just as it had in the Ballinger-Pinchot controversy, Alaska again became the focus of a conflict between the Agriculture Department and conservationists on the one hand, and the Interior Department and private exploiters on the other.

Throughout 1921 there were several proposals before Congress relating to the control of Alaskan resources, but attention centered primarily on three bills. Introduced by Representative Charles F. Curry of California shortly after the new administration took over, the first of these measures, derived from a plan presented to the previous session of Congress, proposed the creation of a local development board to manage the territory's resources. Based on the theory that bureaus in Washington were too far away to supervise effectively the development of Alaska and that officials in the Capital were insensitive to the needs of the territory, the Curry bill would have placed responsibility in the hands of residents who, according to its sponsor, "think in terms of Alaska, and know Alaska, its people, and its needs by personal observation, continuous association, and close contact, and not by theorizing from afar." [7] The proposed board was to take over Alaskan activities currently under the direction of the Departments of Agriculture, Commerce, and the Interior and the Federal Power Commission. Its decisions, however, were to be subject to review by the Secretary of the Interior and ultimately by the President himself.

Wallace and the conservationists, as might well be expected, were unalterably opposed to the Curry bill. The secretary thought that it would be "most unwise and greatly against the public interest . . . [to] turn the forests over to a local board to

[7] Curry to Wallace, May 17, 1921, Files of the Secretary of Agriculture, National Archives.

do with them as it might see fit." In a published letter to Curry, Wallace expressed the traditional conservationist arguments in favor of regulated development of resources under federal supervision. He complained that the legislation in question "would remove wholesome checks against the exploitation of our vast natural resources and would deprive Alaska of the technical help of the Federal agencies which combine long experience and the best scientific knowledge in studying and developing specific natural resources." Calling attention to the fate of timberlands not under federal control, the secretary emphasized the importance of retaining administration of all national forests within the Agriculture Department. Contrary to Curry's charges, he insisted that Alaska's timber development was progressing at a reasonable rate and, more importantly, on a sound basis.[8] Congratulating Wallace on his stand, Pinchot commented that the letter to Curry was "a dandy and ought to settle the issue." [9]

Fall's attitude toward the bill was uncertain. Clearly falling short of the complete control he sought, the provision for review by the Secretary of the Interior would have given him an instrument, although essentially a negative one, for imposing his policies on the development board. The press reported that Fall favored the bill, and Harry A. Slattery, one of the country's most ardent conservationists, wrote to Pinchot that the Interior Department was backing the legislation and perhaps even took part in drafting it.[10] Although Fall later claimed that he had opposed the Curry bill and had favored instead control of Alaskan resources under a single federal agency, the Interior secretary remained silent while Congress considered the measure.[11]

[8] Wallace to Curry, May 14, 1921, *ibid.*

[9] Pinchot to Wallace, May 24, 1921, *ibid.* and Pinchot Papers, Box 243.

[10] *Des Moines Register,* May 21, 1921, p. 6; and Slattery to Pinchot, May 10, 1921, Pinchot Papers, Box 242.

[11] "Notes of G. P.'s talk with Secretary Fall, July 29, 1921," ms. in

Whatever his attitude on the development board scheme, Fall was clearly more interested in a substitute proposal drafted by the Interior Department in May 1921.[12] Embodying his demands in regard to Alaska, this measure was subsequently presented to Congress, after which the Curry bill was permanently shelved. Introduced in August 1921 by Senator Harry S. New of Indiana, a longtime foe of the conservationists, the bill provided for the transfer to the Interior of all functions relating to Alaska currently under the Agriculture Department. Although Harding made no direct statement, the action of Fall and New reportedly had his approval. Placed in an uncomfortable position, the President secretly favored the transfer but wished to avoid taking sides in the cabinet conflict which the bill was certain to provoke. Quietly staying in the background, he allowed Fall to carry on the campaign for the New bill.[13]

Since the Harding administration had taken office, conservationists had been closely watching Fall and the Interior Department. With growing rumors of an imminent effort to transfer the Forest Service, Pinchot gained an interview with Fall in late July 1921 to discuss the matter. Assuring Pinchot that he had no desire to take over the Forest Service, the Interior secretary explained that he only wanted to gain control of government timber sales and grazing rights on the national forests of Alaska in order to coordinate these activities with similar functions of his own department. Despite Fall's denial, Pinchot was convinced by the "whole trend of his conversation that he believed the administration of Alaskan forests should be centered in the Interior Department. . . ."[14] The New Bill removed any doubt about

Pinchot Papers, Box 1897; and Fall to N. J. Sinnott, March 3, 1922, Fall Papers, Correspondence relating to Agriculture and Forests.

[12] Memorandum, E. W. Nelson to Wallace, May 24, 1921, Files of the Secretary of Agriculture.

[13] *New York Times,* July 17, 1921, vol. 7, p. 2.

[14] "Notes of G. P.'s talk with Secretary Fall, July 29, 1921."

Fall's intentions regarding the forest resources of the territory and strongly suggested the possibility of further attempts to bring about the complete transfer of the Forest Service if the measure should pass. Under the leadership of Pinchot, therefore, conservationist forces set out to block the efforts of the Interior Department.[15]

Moving slowly at first, the campaign gained momentum as the designs of Fall and his department became increasingly clear. Harry A. Slattery, a Washington attorney and close friend of Pinchot, opened the attack in September 1921 by arranging for the publication of a series of articles in the *Christian Science Monitor* explaining the function of the Forest Service and outlining the reasons why it should remain in the Agriculture Department.[16] The following month another archconservationist, former Chief Forester Henry S. Graves, placed an article in *American Forestry,* in which he developed a rather elaborate argument in support of the conservationist position. Since forestry was closely related to agricultural development and the national forests served primarily rural communities through grazing privileges and farm land entries, Graves insisted that the Forest Service properly belonged under Wallace's department. With a staff of technical experts to handle the problems of plant life and animal husbandry connected with supervision of the forests, moreover, the Agriculture Department was the logical administrative agency. Graves' most revealing argument, however, was his contention that the attitude of the Department of the Interior toward government lands was conditioned by a "century of disposing of public domain" and that there was no reason to believe that Fall would be inclined to break with tradition and substitute a program of conservation if he gained control of the national forests. In his

[15] Pinchot to A. N. Hume, January 31, 1921, Files of the Secretary of Agriculture.
[16] Slattery to Pinchot, September 22, 1921, Pinchot Papers, Box 242.

concluding shot, Graves held that Agriculture, unlike the Interior, was not influenced by political factors in the formulation of its policies.[17]

In the meantime, Pinchot was enlisting the support of the National Conservation Association, of which he was president and Slattery secretary, the American Forestry Association, and various state forestry groups in the fight against Fall. Preparing a massive petition and press campaign, he hoped to arouse public sentiment to a point where the transfer of any of the Forest Service functions would be politically unthinkable. Obviously enjoying his encounter with Fall and the Interior Department, Pinchot wrote to a friend in October 1921 that "we are really getting to the fascinating stage of the 'fascinating fight.' " [18]

With the introduction of another bill in November by Senator William H. King of Utah, the conservationists stepped up their campaign considerably. Proposing what many of them had been expecting since the previous March, the King measure provided for the complete transfer of the Forest Service and all of its functions to the Interior Department. For some time Fall had been pressuring several congressmen to sponsor such a measure, and the King bill was clearly in response to his request.[19] Although refusing openly to endorse the legislation, the secretary publicly indicated that in the interest of efficiency he favored either the transfer of the forestry division to his department or the placing of Interior Department functions relating broadly to forests in the Agriculture Department.[20] Few, however, were naive enough

[17] *American Forestry* 27 (October, 1921): 645–47.

[18] Pinchot to Arthur F. Fisher, October 17, 1921, Pinchot Papers, Box 237.

[19] Ralph M. Sayre, "Albert Baird Cummins and the Progressive Movement in Iowa," Doctoral dissertation, Columbia University, 1958, p. 535.

[20] *Washington Post*, November 10, 1921, p. 3; and *Annual Report of the Secretary of the Interior*, 1921, pp. 6–7.

to believe that Fall was seriously proposing the latter solution.[21] Thus, the King bill convinced even the most apathetic conservationists of the need for concerted action to preserve the Forest Service in a friendly department.

Pinchot, with the help of Slattery and Graves, redoubled efforts to rally opposition to the New and King bills. He advised the Maine Forestry Association "to leave nothing undone that will make evident to Congress and to the President how firmly . . . [you] are determined not only that the present move shall be defeated, but that it shall be defeated so decisively that it will never be made again." Corresponding with other forestry associations and the Society of American Foresters, Pinchot urged "strong action" against the bills and recommended that the organizations "make known [their] feelings to congressmen." He suggested that "a general protest against the transfer . . . as given in Graves' statement in *American Forestry* would be admirable." From a list provided by the Forest Service, Pinchot wrote numerous individuals and organizations interested in resource development in an effort to enlist wide support for the campaign.[22] On December 23, 1921, the *New York Times* carried a letter from Pinchot in which he called for rejection of the Forest Service transfer and, applying a major piece of conservationist strat-

---

[21] The Interior Department, in fact, drew up a manuscript citing its case for the transfer, which clearly indicated that it had no intentions of giving up claims to the Forest Service much less of relinquishing its responsibilities relative to the national forests to the Agriculture Department. See "Reasons for enactment of S. 4630," Files of the Secretary of the Interior, National Archives, File 2-177, Part 2. Fall later claimed he had always favored transfer of the Forest Service to his department. See "Notes of interview with Fall, November 26, 1921," ms. in Pinchot Papers, Box 1897.

[22] Pinchot to S. T. Dana, December 29, 1921; Pinchot to R. C. Bryant, December 8, 1921; telegram, Pinchot to Harris A. Reynolds, December 8, 1921; Pinchot to Philip P. Wells, December 8, 1921; Pinchot to Fred Brenckman, undated; and Pinchot to E. A. Shearman [*sic*], November 19, 1921, Pinchot Papers, Boxes 236, 237, 241, 243.

egy, charged that "the same interests that attempted to gobble up the natural resources of Alaska in Secretary Ballinger's time are at it again." [23] Sent out to some 5,000 newspapers across the country and to every forestry association, the letter became a familiar document in the campaign. Securing a meeting with Harding, Pinchot also sought to persuade the President of the advisability of leaving the Forest Service in the Agriculture Department. "I told him [Harding]," Pinchot wrote of the interview, "that the result of the proposed transfer would be to abolish the results of my life's work." But the President was apparently unmoved by this plea, for he refused to commit himself.[24]

The campaign of the conservationists resulted in another meeting between Pinchot and Fall in late November 1921, this time at Fall's request. Quietly summoning Pinchot, the Interior secretary tried to convince him to change his mind on the Forest Service question. Contrary to what Fall had indicated in their earlier interview, he now maintained that he had always favored the transfer and attempted to show the former Chief Forester the logic of his position. Pinchot of course remained unconvinced, and the discussion accomplished nothing except to demonstrate to the conservationists that their efforts were beginning to take effect.[25]

As public interest was aroused over the forestry question, letters began to flow into the Departments of the Interior and Agriculture protesting against or supporting the proposed transfer. United in opposition to the New and King bills were the major farm groups, various forestry organizations, college forestry departments, and the Agricultural Editors Association. On the other hand, mining interests and electrical power companies gen-

[23] *New York Times,* December 23, 1921, p. 23.
[24] "Gifford Pinchot's interview with President Harding, November 29, 1921," ms. in Pinchot Papers, Box 1897.
[25] "Notes of interview with Fall, November 26, 1921."

erally favored the transfer of the Forest Service to the Department of the Interior. Lumbering concerns and stock grazers were divided, as groups holding privileges in the national forests favored Agriculture jurisdiction and those without tended to back the Interior. On the average, however, sentiment ran strongly in opposition to the transfer.[26] So impressed was Pinchot with the response to the campaign that he confided to his brother in late December that the conservationists appeared to have won their fight.[27] And well he might have been pleased, for the New and King bills found little support in Congress and neither was reported out of committee.

Actively joining in the campaign, the influential American Forestry Association issued a propaganda sheet in December 1921 which shortly became a point of controversy between the Departments of the Interior and Agriculture. Sent out to various newspapers and to association members, the flier contained several articles opposing the transfer of the Forest Service or any of its functions in regard to Alaska. Urging members to "take this news sheet to the editor of your newspaper and impress upon him its importance," the association fully committed itself to the fight against Fall. The item which attracted the most attention was a partial reprint of an article by William B. Greeley, currently Chief Forester, which had appeared in the April issue of the association's journal, *American Forestry,* and in which he had attacked the Alaskan development board scheme.[28] Although the article had been reprinted without the Chief Forester's consent or prior knowledge, its appearance in the flier gave

[26] See Files of the Secretary of Agriculture; Files of the Secretary of the Interior, File 2-177, Part 2; and Herbert C. Hoover Papers, Herbert Hoover Presidential Library, West Branch, Iowa, 1-I/258.

[27] Pinchot to Amos Pinchot, December 23, 1921, Pinchot Papers, Box 241.

[28] Copies of the flier in Files of the Secretary of Agriculture and Harding Papers, Box 23. For Greeley's complete article see *American Forestry* 28 (April, 1921): 200–209.

the impression that the Agriculture Department was actively assisting in the conservationists' propaganda campaign.[29]

Upon learning of the American Forestry news sheet, Fall, who was out of Washington at the time, wired his office to "call President's attention to Greeley matter." He recommended that the department issue a press release to the effect that official comment on matters pending before Congress should be made through government channels in order "not to bias or prejudice or influence legislation by propaganda and untruthful statements." The Interior chief thought it best, however, to avoid specific reference to Greeley or the Agriculture Department. On the following day Fall's office sent Harding a copy of the flier and issued a statement to the press, and for a time the matter went no further.[30]

The principal reason why the conflict between the two departments did not flare into the open sooner was Wallace's refusal to be drawn into the fight in its early stages. After his strong statement against the Curry bill, he remained silent and let the conservationists carry on the campaign against the New and King bills. This strategy caused some resentment in the conservationist camp. "Unless Secretary Wallace steams up some," Slattery complained early in the campaign, "he will find it too late. . . ."[31] Pinchot, who was "perfectly sick" that the Forest Service had not used its influence among the friends of conservation to better advantage, made an appointment with Wallace "to go over the whole thing." The meeting must have satisfied him, for despite the fact that "Greeley had been letting the whole thing go by de-

[29] Greeley to Wallace, March 6, 1922, Files of the Secretary of Agriculture.

[30] Telegram, Fall to E. C. Finney, December 30, 1921; E. C. Finney to Harding, December 31, 1921; and Department of the Interior press release, December 31, 1921, Files of the Secretary of the Interior, File 2-177, Part 1.

[31] Slattery to Pinchot, September 14, 1921, Pinchot Papers, Box 242.

fault," he conceded a short time later that "Wallace has at last got his fighting clothes on. . . ."[32] Pinchot's declaration was premature, however, for it was to be three months before the secretary publicly revealed his intention to oppose his colleague in the cabinet.

Given Wallace's querulous temperament and his readiness to defend his prerogatives, his hesitancy in entering the fight against Fall was difficult for many conservationists to understand. Certainly one reason for the delay was a reluctance to precipitate an interdepartmental quarrel if it could possibly be avoided. The secretary felt no urgency in the matter, furthermore, because Harding had promised to discuss the transfer question with the cabinet before sanctioning any action either by Congress or the Joint Committee on Reorganization.[33] Thus Wallace decided to wait and, if he had to commit himself at all, to pick the most advantageous time to join the fray. Despite Pinchot's comments to the contrary, Greeley was most impatient and offered to resign so that he might carry on the fight outside of the department. Wallace refused to accept the resignation, however, and succinctly, if somewhat crudely, explained his strategy to the impetuous Chief Forester: "My Boy, don't ever get yourself into a pissin' contest with a skunk. This is my fight as much as yours. We'll take it on; but let's pick our own ground. Get the sun in the other fellow's eyes."[34]

As the conservationist campaign gained momentum, Fall's attacks on the Forest Service became more pointed and his actions less prudent.[35] In late January 1922 he gave an interview to the

[32] Pinchot to P. S. Lovejoy, November 18, 1921; and Pinchot to F. E. Olmsted, December 12, 1921, *ibid.,* Boxes 239, 240.

[33] Harding to Albert B. Cummins, November 10, 1921, Albert B. Cummins Papers, Iowa State Department of History and Archives, Des Moines, Box 24.

[34] Greeley, *Forests and Men,* p. 99.

[35] Pinchot addressed the National Agricultural Conference and used the opportunity to speak out against the transfer of the Forest Service.

El Paso Herald in which he directly criticized the Agriculture Department's administration of the national forests. Ignoring what he had said only a short time earlier regarding the possible transfer of overlapping functions to the Department of Agriculture, Fall now flatly stated that the Forest Service belonged under his jurisdiction.[36] In addition, the Interior Department engaged a certain Haviland H. Lund to collect information to strengthen its position in the fight. The purpose was apparently to discredit Wallace and Pinchot, for in February Lund reported: "Regarding Mr. Wallace, I am not overlooking your request in that direction, and think I will have something definite to say within a week." Although there is no evidence that he found anything that might be used against Wallace, Lund suggested a short time later an "anti-Pinchot program," through which "his influence can be entirely annihilated. . . ." The Interior Department's investigator had allegedly uncovered documents proving that Pinchot was the "advance guard of the German propaganda instigated here many years before the War, with a view to breaking down our Government." By "attaching the word 'German' to him," Lund advised Fall, "you can easily turn the tables." [37] This scheme must have been too much for even the unscrupulous Interior secretary, however, for he failed to follow up Lund's suggestion.

Finally, in early March 1922, Fall went beyond criticism of

Slattery launched a scathing attack on Fall in a speech to the National Popular Government League in March 1922. In addition, the press and petition campaign continued into 1922. United States Congress, House, *Report of the National Agricultural Conference,* 1922, Document 195, 67th Cong., 2nd Sess., p. 111; address by Slattery, March 10, 1922, copy in files of the Secretary of Agriculture; *American Forestry* 29 (January, 1922): 25, 31, 48–49 and 29 (March, 1922): 165.

[36] *El Paso Herald,* January 24, 1922, pp. 1–2.

[37] Memorandum, Lund to C. V. Safford, February 9, 1922; and Lund to Fall, March 14, 1922, Files of the Secretary of the Interior, File 2-177, Part 2.

the administration of the Forest Service and began attacking personalities and calling into question the motives of his opponents. In a published letter to the chairman of the House Committee on Public Lands, Nicholas J. Sinnott of Oregon, he reduced the discussion on the transfer to a diatribe against the Agriculture Department and virtually forced Wallace to respond. Referring to the excerpts from Greeley's article in the American Forestry news sheet, Fall wrote that his office was "outraged at this vicious and unwarranted attack upon the head of a coordinate department of the Government . . . by the chief forester of a bureau in the Department of Agriculture." He charged that "certain narrow-minded and biased bureaucratic government officials and their sponsors" were responsible for the "stupid and arbitrary regulation of the Forestry Bureau," which had severely retarded the development of Alaska. Either private interests had to have the "speculative chance" denied by the Agriculture Department's policy, Fall maintained, or the government itself would have to assume the risks of developing Alaskan resources. If the Forest Service were under his control, the Interior secretary guaranteed, that chance would be provided without opening the territory to wasteful exploitation, thereby precluding the need for federal development. Posing as a "follower of Lincoln" from whose ideas on resource development he claimed to have drawn, Fall was confident that his policy rather than "Pinchotism" or "Greeleyism" was the rational way to administer the public domain.[38]

In his published reply Wallace was terse and to the point. As for Fall's contention that he had been subjected to a vicious attack, the Agriculture secretary declared, there was "absolutely no foundation for such a charge." Pointing out that Greeley had prepared his article in the fall of 1920 and that it had gone to

[38] Fall to Sinnott, March 3, 1922, Fall Papers, Correspondence relating to Agriculture and Forests; and Department of the Interior press release, March 4, 1922, copies in Files of the Secretary of the Interior, File 2-177, Part 2 and Files of the Secretary of Agriculture.

press no later than March 15, 1921, less than two weeks after the new administration had taken office, Wallace argued that it could not be considered an attack on Fall. The timing of the publication notwithstanding, it merely expressed the Agriculture Department's well-known opposition to the Alaskan development board scheme and was in no way meant to criticize the activities of another department. In regard to the appearance of the article in the American Forestry flier, the secretary explained that it was without Greeley's consent or previous knowledge. Wallace refused to make a statement on the Forest Service transfer, however, choosing to withhold comment as long as the matter was pending.[39] Although his reply was perhaps not as strong as the conservationists would have wished, they nevertheless welcomed the secretary into the fight against Fall and the Interior Department.

If Wallace followed a policy of moderation in public utterances, his conduct within administration circles was much less guarded. Hoping to prevent an open conflict between executive departments, Harding consciously discouraged discussion of the transfer among his advisors. As time passed and nothing was resolved, however, Wallace became impatient and finally brought up the matter at one of the cabinet meetings. A bitter argument ensued in which the Agriculture secretary threatened to resign and take his case to the public if the Forest Service were transferred.[40] About the same time the Press reported that Wallace was interested in leaving the Agriculture Department for the new position recently created on the Federal Reserve Board. Though he eventually denied this, the secretary delayed long

[39] Wallace to Arthur Capper, March 11, 1922; USDA press release of the letter, March 12, 1922; copies of the letter sent to members of the Committees on Agriculture and Forestry, Public Lands, and Territories in both Senate and House; Wallace to Fall, March 16, 1922; Wallace to Fall, March 23, 1922; and Greeley to Wallace, March 6, 1922, Files of the Secretary of Agriculture.

[40] Alfred Lief, *Democracy's Norris* (New York: Stackpole Sons, 1939), p. 264; and *Wallaces' Farmer* 49 (October 31, 1924): 1420.

enough to allow a flood of protests to reach the White House.[41]

After March 1922 the Forest Service controversy rapidly subsided. For one thing, Fall issued no further statements on the matter. According to the press, he had lost his influence with the President because of public reaction against the proposed transfer and thus given up efforts to gain control of the national forests. In fact, several newspapers suggested that his resignation might come any day.[42] Fall's waning interest in the transfer question, however, was more likely the result of a growing preoccupation with another project, a project which was eventually to lead to his downfall. Having secured control of the Navy's oil reserves he was in the process of illegally leasing them to private interests.

The conservationists' campaign likewise died down. Pinchot's candidacy for the governorship of Pennsylvania, which prevented him from devoting much time to the movement, was certainly one reason. But more important was the fact that there was nothing left to fight. Due in large part to the public sentiment generated by the conservationists, not one bill affecting administration of the national forests was introduced into the long second session of the Sixty-seventh Congress. With Fall silent and the appearance of no further threats to the Forest Service, the conservationists had virtually won their case and there was no need to continue the campaign. There were rumors toward the end of 1922 that the attempt to transfer the national forests would be renewed in the third session of the Sixty-seventh Congress, but they failed to materialize.[43]

[41] *Des Moines Register,* May 26, 1922, p. 1 and June 6, 1922, p. 1; and Wallace to Verne Marshall, June 6, 1922, Files of the Secretary of Agriculture.

[42] *New York Times,* March 12, 1922, vol. 7, p. 1; *Des Moines Register,* May 25, 1922, p. 1.

[43] H. H. Chapman to Wallace, November 3, 1922; Wallace to Chapman, November 6, 1922, Files of the Secretary of Agriculture; and Pinchot to Harding, November 15, 1922, Harding Papers, Box 701.

Despite newspaper accounts to the contrary, the subsiding of the Forest Service question was not due to a change in attitude on the part of Harding. At least as late as November 1922 he still favored control of the national forests by the Interior Department, but not at the cost of conflict within his administration. "I should have been glad long ago," he admitted to Pinchot, "to have submitted an accepted [sic] program [of reorganization] to the Congress, but the lack of harmony among Cabinet heads has made me hesitate to submit a program which offers the prospect of long drawn-out contention." He indicated a reluctance to endorse any plan without a "measurable satisfactory agreement" among his advisors. Harding chose, therefore, to let the matter rest and merely to renew his assurance of a full cabinet discussion before settling on a reorganization program.[44]

But the reorganization question could not be ignored indefinitely. Despite the fact that Congress had dropped the Forest Service issue, the Joint Committee on Reorganization continued to function and was committed to offer some kind of recommendations. Harding had originally anticipated that the committee's plan would be ready to place into operation in the fiscal year beginning July 1, 1922, but the fight over the Forest Service had delayed progress.[45] Conferring with Brown in July 1921, Wallace learned that the committee chairman intended to recommend Interior control of the forests in his report to Harding. In a statement to Brown a short time later, the Agriculture secretary denounced the proposed transfer and suggested that it be dropped. Wallace also urged opponents of the move to make their views

[44] Harding to Pinchot, November 18, 1922, Harding Papers, Box 701; Wallace to Verne Marshall, June 6, 1922; and Wallace to H. H. Chapman, November 6, 1922, Files of the Secretary of Agriculture.

[45] Charles G. Dawes, *The First Year of the Budget of the United States* (New York: Harper and Brothers, 1923), pp. 5–6.

known to the joint committee.[46] But the committee could not be influenced, and its first tentative report to the President in early 1922 recommended placing the Forest Service under the Department of the Interior. Harding quietly sent the report back for reconsideration without presenting it to the cabinet. A second report a short time later was no more satisfactory, and it too was summarily rejected by the President.[47]

Dissatisfied with the progress of the joint committee, Harding finally instructed the cabinet in late 1922 to take the initiative in the reorganization question. Evidently the discussions proceeded satisfactorily, at least in Wallace's opinion, for he could report in December that "there is some hopefulness in prospect." [48] After several weeks' consideration, the cabinet submitted a list of suggestions to guide the work of the Joint Committee on Reorganization. Sent to the committee in February 1923, the report contained only two recommendations affecting Wallace's department—the transfer of the Bureau of Roads from Agriculture to the Interior, to which the secretary reluctantly agreed, and the transfer of the Botanic Gardens from congressional supervision to control of the Agriculture Department. Interestingly, no mention was made of the Forest Service. "In a few instances," Harding had to admit, ". . . the principle or major purpose has not been followed to the letter, in order to avoid controversies which might jeopardize reorganization as a whole," but he felt these

[46] Wallace to Harding, August 8, 1921; and Wallace to Brown, January 16, 1923, Files of the Secretary of Agriculture.

[47] There is no record of the first report, and its exact date is uncertain. The *Washington Herald* and the *Washington Post* placed it in early January. It could have been earlier, however, for Slattery wrote to Pinchot in November 1921 that he had heard that "without question Harding is very much disappointed in Brown's work. . . ." *Washington Herald,* March 11, 1922, p. 1; *Washington Post,* January 22, 1922, p. 2; and Slattery to Pinchot, November 4, 1921, Pinchot Papers, Box 242.

[48] Wallace to A. R. Mann, December 30, 1922, Files of the Secretary of Agriculture.

were of "minor importance." [49] The joint committee was to study the matter for yet another year and finally to submit a proposed bill which followed closely the cabinet's suggestions in regard to executive reorganization. At the hearings on the legislation in April 1924, Wallace climaxed the long, involved episode with a rather comical twist by recommending that the entire public domain be placed under the Agriculture Department, where "its use may be so regulated as to check the present destructive processes." [50] His suggestion was ignored, however, and reorganization proceeded along the lines set down by the committee.

At the time the cabinet was considering the matter of reorganization the announcement of Fall's resignation came out. Press reports in December 1922 indicated that the Interior secretary intended to quit and again speculated that his decision had resulted from a loss of influence with Harding. Early in the next year the White House confirmed that Fall was resigning, effective March 4, 1923, but emphasized that it had nothing to do with reported differences over reorganization. Rather, the Interior secretary was supposedly leaving the cabinet to devote full time to his business interests.[51] Besides the rebuff on the Forest Service issue, Fall's leasing of government oil reserves was under senatorial investigation. In view of the fact that this inquiry soon uncovered irregularities that eventually led to his conviction and imprison-

[49] United States Congress, Senate, *Reorganization of the Executive Departments,* Document 302, 67th Cong., 4th Sess., pt. iii, pp. 1–2. Wallace indicated to Harding in August 1921 that he had no objection to transfer of the Bureau of Roads from Agriculture to Interior. Wallace to Harding, August 8, 1921, Files of the Secretary of Agriculture.

[50] United States Congress, Joint Committee on Reorganization of the Administrative Branch, *Hearings on Reorganization of the Executive Departments,* 68th Cong., 1st Sess., pp. 277–79. Wallace had earlier indicated to Taylor his desire that the reclamation division, national parks, and administration of all grazing on the public domain be transferred to Agriculture. Wallace to Taylor, November 30, 1923, Henry C. Taylor Papers, State Historical Society of Winconsin, Madison.

[51] *Washington Post,* January 3, 1923, p. 3 and January 4, 1923, p. 6.

ment for accepting bribery, it undoubtedly was the major factor in his decision to resign. Whatever his reasons, Fall's departure removed the last threat to the Forest Service, for his successor, the genial Hubert Work, had little interest in the transfer.

The Forest Service safely in the Agriculture Department, it remained to make a conservationist of Harding. Not only was the President still convinced that the transfer would have improved administrative efficiency, but he remained unimpressed with the talk of scientific and regulated resource development. Since Wallace and Greeley favored the expansion of forest reserves and a broader federal forestry program, they were naturally anxious to gain his sympathy and support. The Agriculture secretary therefore decided to take advantage of Harding's western tour in mid-1923 to give the President an education in conservation.

Accompanied by an entourage including Hoover, Work, and later Wallace, who joined the party en route, Harding left on June 20, 1923. The group was scheduled to visit the Great Plains, the Northwest, and Alaska to observe agricultural conditions and to get the reactions of the people to the administration's farm program. But the tour also offered an opportunity, as Wallace fully realized, for the President to observe Alaskan developments firsthand and to assess the results of the Forest Service program. About a month before the trip the secretary had written to Harding urging him to familiarize himself with the work of the Agriculture Department in Alaska.[52] Wallace was personally pleased with the opportunity to make the tour, for it gave him a chance to talk with Alaskans who felt that the department's policy was retarding resource development in the territory.

The climax of the trip, at least for Wallace and the conservationists, came with Harding's speech at Seattle, Washington, on

[52] Wallace to Harding, May 26, 1923, Files of the Secretary of Agriculture.

July 27, 1923. Having just returned from Alaska, the official party was in the last stages of the tour. In his Seattle address, Harding surprised nearly everyone by reversing his stand and adopting the conservationist line. "I must confess," he admitted, "I journeyed to Alaska with the impression that our forest conservation was too drastic and that . . . protests would be heard on every side, . . . [but] I had a wrong impression. . . ." Wondering out loud if opponents of the Forest Service policy had been to Alaska, Harding ventured that their concern was not inspired by an interest in Alaskan development as much as by "the hope of getting . . . [the territory] turned over to wholesale exploitation. . . ." Although there had been some instances of "excessive restrictions," the President was satisfied that the overall program for development of the territory was the "safe plan." "There is no broad 'problem of Alaska,' " he concluded, "despite insistence on its existence." [53]

This was more than conservationists had ever hoped. Elated with the Seattle speech, Wallace thought that it would be remembered as "one of the greatest of President Harding's public utterances." Despite the fact that it dealt specifically with Alaska, he felt that the address had laid down "certain fundamental principles" which applied generally to resource development.[54] Pinchot sent Wallace "vigorous congratulations . . . on the magnificent result of your trip to Alaska." Pleased that "the good old Roosevelt Alaska policy has been sustained," he was confident that "Fall and all his work so far as Alaska is concerned has [sic] been definitely and finally repudiated." Pinchot was satisfied, moreover, that Harding's speech had removed the possibility of any further efforts to transfer the Forest Service.[55]

[53] Address on the Territory of Alaska at Seattle, Washington, July 27, 1923, Harding Papers, Box 152.

[54] USDA press release, August 31, 1923, copy in files of the Secretary of Agriculture.

[55] Pinchot to Wallace, August 2, 1923, Pinchot Papers, Box 254.

Harding's change of heart was difficult to account for. Convinced that Wallace had finally won him over, Pinchot wrote to his friend: "I take it that the President's speech at Seattle was directly due to you." [56] Greeley, who also made the western trip, and Taylor likewise attributed Harding's reversal to the influence of their chief.[57] On the other hand, Wallace explained the change as the result of the President's seeing Alaskan conditions in person, which demonstrated to him that opponents of the Forest Service policy had misrepresented the facts. "He [Harding] started to Alaska with many wrong impressions," the secretary wrote to his family, ". . . [and] was not at all in sympathy with my views. . . ." But the more he observed of the territory and talked with its people, Wallace related, "the more clearly he saw that he had been fooled . . . ," with the result that "he fully endorced [sic] my policies." [58]

A more likely factor in Harding's sudden conversion to conservationism than either Wallace's influence or the revelation of the Alaskan trip was the discovery, or at least the suspicion, of the scandal surrounding Fall. Calvin Coolidge later revealed that before leaving on the western tour the President had learned that "someone whom he had trusted betrayed him. . . ." [59] During the stop at Kansas City, Fall's wife talked with Harding, which according to William Allen White, left him "perplexed and perturbed." On the following day the President supposedly made his classic remark to the Emporia editor: "I have no trouble with my enemies. . . . It is my friends that are giving me trouble." White also claimed that Harding received a coded message while in

[56] *Ibid.*

[57] Greeley, *Forests and Men,* pp. 100–101; and Henry C. Taylor, "A Farm Economist in Washington," pp. 251–52, ms. in Taylor Papers.

[58] Wallace to Folks, August 4, 1923, Henry C. Wallace Papers, State University of Iowa Library, Iowa City, Box 3.

[59] Calvin Coolidge, *The Autobiography of Calvin Coolidge* (New York: Cosmopolitan Book Corporation, 1929), p. 168.

Alaska which caused him considerable distress.[60] Perhaps Wallace was alluding to these developments when he wrote to his family from Alaska that "the work of the Department of Agriculture has grown steadily in the favor of the newspaper men and others of the party, while some other sorts of work have not been so fortunate." [61] At any rate, the President's Seattle speech came after he had some indication of Fall's betrayal, and there was no doubt a connection between his change of attitude and the information passed to him. Certainly parts of the address would suggest that he had experienced something more shocking than Wallace's persistent conservationism or a quick tour of Alaska. The switch in position was too abrupt and the tone too bitter to be explained so simply.

The time now seemed right for the conservationists to try once again to gain the expansion of the federal forestry program they had been urging since the beginning of Harding's administration. Throughout 1921 the House Committee on Agriculture and Forestry, in cooperation with the Department of Agriculture, had worked on legislation along this line. In early 1922 the committee held hearings on the subject, at which appeared Wallace, Greeley, Pinchot, Graves, and several others interested in conservation. After the hearings the committee, with Greeley's help, drafted a bill providing for the extension of forest reserves and federal-state cooperation in the formulation of a national forestry program. In February 1923 a member of the House committee, John D. Clarke of New York, finally introduced the measure into Congress. Several times in the previous two years Wallace had sought Harding's endorsement of a conservation policy along

[60] William Allen White, *A Puritan in Babylon* (New York: Macmillan, 1938), pp. 238–39; Walter Johnson, *William Allen White's America* (New York: Henry Holt, 1947), p. 373; and Francis Russell, *The Shadow of Blooming Grove: Warren G. Harding in His Times* (New York: McGraw-Hill, 1968), pp. 573–89.

[61] Wallace to Folks, July 23, 1923, Wallace Papers, Box 3.

the lines of the Clarke bill, but, though the President expressed general agreement with the objectives of the measure, he refused to give it his support.[62] As a result, despite the backing of Wallace, the Forest Service, and the conservationists, the legislation generated little interest in the Sixty-seventh Congress. With the President's change of heart toward the Forest Service work in Alaska, the prospects for expanding the entire forestry program appeared to be measurably improved.

The optimism of the conservationists was short-lived, however, for six days after his Seattle speech Harding died and no one was sure of the attitude of incoming President Calvin Coolidge. Still, a hard core of support for change in the forestry policy remained and was ready to push its program when the Sixty-eighth Congress opened in December 1923. Shortly after the legislators convened, Senator Charles L. McNary of Oregon introduced a bill along the lines of Clarke's, and the following month Clarke reintroduced his own measure. In its final form the Clarke-McNary bill included provisions for the expansion of national forests and federal-state cooperation in reforestation and fire-prevention programs, both of which Wallace had been advocating since he took office. Insisting that such legislation was necessary to insure the country an adequate timber supply in the future, the secretary, with little success, strongly urged Coolidge's support.[63] Largely due to the influence of Wallace and the work of the conservationists, Congress passed the Clarke-McNary bill in June 1924. Greeley declared that the new law represented

[62] Wallace to Harding, November 21, 1921, Harding Papers, Box 203; Wallace to Harding, December 3, 1922; Harding to John D. Clarke, January 24, 1923, Files of the Secretary of Agriculture; and "Report of the Secretary," USDA *Yearbook of Agriculture,* 1922, p. 36.

[63] Wallace to Coolidge, January 10, 1924, Calvin Coolidge Papers, Manuscript Division, Library of Congress, File 149, Box 125; and *New York Times,* April 20, 1924, vol. 11, p. 18.

"the greatest forward step in the forest policy of the United States since the Weeks law of 1911." [64]

Thus conservationists not only succeeded in turning back an attempt by the Interior Department to capture the Forest Service, but also managed to expand the forestry program of the federal government. In both instances Secretary Wallace was a key figure.

Wallace's role in opposing the efforts of Fall and the Interior Department and in expanding forestry work was in many respects one of his most significant as Secretary of Agriculture. Although he was slow to enter the fight, his part was probably more important in preventing the Forest Service transfer than that of other conservationists. While the contributions of Pinchot, Slattery, Graves, and others were valuable, Wallace's steadfast resistance in the cabinet no doubt had the greater influence on the President. Had the secretary agreed or at least acquiesced in the Forest Service transfer, Harding might well have risked offending public opinion and persisted in pushing the desired reorganization. Not once did the President cite popular opposition as a factor in either his reluctance to move faster or his decision to delete the transfer proposal from the cabinet recommendations to the Joint Committee on Reorganization; rather, the desire to avoid interdepartmental conflict was clearly the primary consideration. And while it is true that legislation accomplishing the transfer of the Forest Service might have been defeated in any event as a result of opposition of conservationists in and out of Congress, the full backing of the President would have enhanced its chances greatly.

Fall, of course, might have been truly interested in initiating a

[64] USDA press release, June 23, 1924, copy in files of the Secretary of Agriculture. The Weeks law provided for federal-state cooperation in the protection of forested watersheds of navigable rivers.

program of rational use of national forest resources and sincere in his desire to improve the administration of the Forest Service. The Department of Agriculture, it was true, moved cautiously in opening the forests for private timber and wood pulp development, grazing, and land claims. In the immediate postwar period Alaska was suffering a steady decline in population and severe economic problems. Due largely to the exhaustion of its gold mines, which had sustained a territorial boom since 1896, and to the general worldwide depression, these conditions might have been partially relieved by a more vigorous program of resource development.[65] In criticizing the reluctance of the Department of Agriculture to open all timberlands to private groups, furthermore, Fall argued with more than a little logic that it was as wasteful to prevent use as it was to over-use natural resources. Under his control, the Interior secretary guaranteed, the national forests would be made to serve the needs of the country.

Although Fall might have been interested in resource development, it is highly doubtful. His handling of the naval oil reserves, which were turned over to private exploiters in return for bribes, demonstrated the kind of program he probably had in mind for the national forests. While they could not be proved, there were rumors at the time that the Morgan-Guggenheim interests, working through the Seattle Chamber of Commerce, were behind the efforts to transfer the Forest Service.[66] These same interests had

[65] For the Forest Service program see Wallace to C. F. Curry, March 14, 1921; Greeley to Wallace, January 5, 1922; "Achievements of the Forest Service," ms. in files of the Secretary of Agriculture; "Why the Forest Service Should Remain in the Department of Agriculture," ms. in Harding Papers; and Clinton G. Smith, *Regional Development of Pulpwood Resources of the Tongass National Forest, Alaska,* USDA Bulletin 950 (1921): 4.

[66] Pinchot to John H. Cobb, January 5, 1921, Pinchot Papers, Box 237; John E. Ballaine to Wallace, undated; Ballaine to Hubert Work, April 12, 1923, Files of the Secretary of Agriculture; Ballaine to C. Bascom Slemp, September 28, 1923, Coolidge Papers, File 400B, Box 176; and *Washington Post,* July 16, 1923, p. 3.

been involved in the Ballinger-Pinchot affair, and the rumors might well have originated with the conservationists in their desire to relate the proposed transfer to the earlier conflict. Whether the Morgan-Guggenheim associates were the instigators and force behind the proposed transfer, however, was really immaterial. The fact remains that this group, which dominated the Alaskan economy, would have benefited greatly from relaxed restrictions on government-owned resources in the territory. Once in control of the national forests, Fall would certainly have been willing, as he was in the matter of the oil leases, to deal with and make concessions to the Morgan-Guggenheim interests or any other party in return for financial considerations for himself.

Wallace's efforts to preserve the Forest Service in the Agriculture Department, therefore, very likely saved the Harding administration from still another scandal. This is not to say that he foresaw Fall's duplicity, but his stubborn refusal to accept the transfer denied the Interior chief further opportunity for deceit. More importantly, it is quite possible that Wallace was instrumental in preventing the country's forest reserves from falling completely into the hands of private exploiters.

Chapter IX / WALLACE
AND REGULATION

CENTRAL to Wallace's solution to the postwar farm crisis and to his plan for a permanent federal agricultural program was the idea of regulation. The work of the BAE in agricultural economics, to be sure, might enable farmers to gain a more favorable competitive position, but in many instances business consolidations had grown to the point where their power and control over the economy was beyond challenge by individual producers or even by their organizations. Not opposed to bigness as such, Wallace believed that large combinations held out the possibility of greater efficiency and thus of benefit to the country as a whole. An inevitable result of a maturing industrial economy, moreover, consolidation had become a fact of American life with which the country had to reckon. Although willing to accept the movement in general, Wallace was quick to condemn the misuse of economic power by big business to gain unfair advantage and retain for itself all of the fruits of improved and expanded organization. But he was confident that this problem could be corrected through federal regulation. Under a program of government supervision, he held, the abusive use of power by large combinations might be checked and at the same time the advantages of consolidation extended to the general public.

As a representative of agriculture, Wallace was of course pri-

190

marily concerned with the regulation of industries upon which farmers were heavily dependent. Since one of the most important of these industries was the railroad, he had for a long while been a staunch advocate of federal intervention to protect the interests of shippers. Founded in 1905 largely through Wallace's efforts and dominated by him before he became Secretary of Agriculture, the Corn Belt Meat Producers' Association, as already pointed out, had as its major objective the securing of more equitable rates and better transportation conditions for Iowa farmers. The organization's main line of attack was to argue the shippers' case before federal and state regulatory agencies and to demand stronger government supervision of the nation's carriers. Despite its unpopularity in many agrarian quarters, furthermore, Wallace had supported the Transportation Act of 1920 principally because the measure provided for closer regulation of the railroads by the Interstate Commerce Commission. It was natural, then, that he should carry his interest in federal regulation of transportation with him when he entered the Harding cabinet.

Concerned over the increase in railway freight rates in the postwar period, Secretary Wallace felt that the advances allowed by the Interstate Commerce Commission on farm products were disproportionately large compared to those on other commodities. Besides discriminating against American farmers in general, they worked a particular hardship on specific agricultural regions of the country. The secretary explained that agriculture had developed and expanded on the basis of low freight rates and with the expectation that transportation costs would, if anything, be reduced as traffic grew. Through minimal long-haul charges, railroads had encouraged the westward extension of agriculture and thus the production of farm products at considerable distances from processing centers and the major consuming markets. "It necessarily follows," Wallace asserted, "that the character of the farming and the value of the land and improvements were determined by this policy, and that any marked change . . . is bound to make necessary profound changes in both agriculture and industry." [1] He believed that the postwar rate increase threatened just such a dislocation, for "the simple process of marking-up the transportation cost a few cents per hundred pounds has the same effect on a surplus-producing State as picking it up and setting it down 100 to 300 miles farther from the market." [2] Western wheat states were a case in point. Faced with both higher freight charges and declining prices, farmers in these areas found themselves at a disadvantage in competition with producers nearer markets and those in other countries using cheaper water transportation. As alternatives there seemed only bankruptcy or a long period of painful readjustment in production and operations to meet the new conditions. The plight of western wheat farmers, moreover, was typical of that of producers in other areas. Unless some adjustment were made in transportation costs, Wallace expected a "rearrangement of production," which augured ill not only for agriculture but for the country as a whole.[3]

The government, to Wallace's thinking, was in part responsible for the farmers' transportation problems. While railroads were under federal control during the war, operational costs increased without check and, as in the case of wages, sometimes with the sanction of the government. The wartime budget absorbed the rising costs of running the roads, however, and rates remained fairly steady during the period of federal operation. But when the lines were returned to private control, Wallace

[1] Address before the New York Academy of Political Science, New York, April 28, 1922, copy in Henry C. Taylor Papers, State Historical Society of Wisconsin, Madison.

[2] "Report of the Secretary," USDA *Yearbook of Agriculture*, 1921, p. 9.

[3] *Ibid.;* Address before the New York Academy of Political Science, New York, April 28, 1922; and Henry C. Wallace, *The Wheat Situation: A Report to the President* (Washington, 1923), pp. 20–24.

maintained, owners could claim that substantial advances in freight charges were necessary, an argument subsequently accepted by the Interstate Commerce Commission. Thus he reasoned that it was incumbent upon the government to provide relief to farmers who were suffering as a result of these increases.[4]

This, of course, did not mean federal ownership or even subsidization, but careful analysis of existing rate structures in relation to agricultural conditions and adjustment to place transportation charges more nearly in line with the producers' ability to pay. Shortly after taking office, Wallace directed Taylor to study freight costs on farm products and to devise, if possible, a formula for bringing about an equitable adjustment. The bureau chief recommended a general lowering of the price level, particularly on foodstuffs, and a corresponding reduction in railway wages in proportion to the decrease in the cost of living, which would in turn permit a cut in rate charges to producers. In this way, Taylor theorized, "the farmers' purchasing power may be in some measure restored, which . . . will enable him to buy the industrial products of the city and set up the normal flow of traffic, all of which will result in increasing the prosperity of the farmers, the railways, and the industrial workers." [5]

The major weakness of Taylor's proposal, as Wallace viewed it, was that it would have required government price controls, a practice to which the secretary was unalterably opposed. In addition, he was little attracted to deflation as a solution to agricultural problems because of the burden it would place on debtor farmers. But there was another solution, one which seemed to Wallace better calculated to meet agrarian needs. "The present rates," he explained, "were all right for the farmer when he had

---

[4] Wallace, *The Wheat Situation*, p. 21; "Notes of Cabinet Meeting, September 25, 1923," copy in Taylor Papers; and *New York Times*, September 26, 1923, pp. 1, 4.

[5] Memorandum, Taylor to Wallace, August 8, 1921, Taylor Papers.

$1.50 corn and hogs at $18 to $20 a hundred pounds, but with corn at 30 cents a bushel and hogs at $8 the freight charges are out of proportion." To Wallace the remedy was obvious—preferential rates for agricultural products. Since farm commodities were basic necessities, he judged, it was only fair that while their prices remained depressed, transportation charges on them should be reduced and the difference made up on less essential goods. He further suggested that it would be in the interest of the country in general to aid farmers in this way, for the health of the national economy was dependent upon the recovery of the agricultural industry.[6] Supporting the former editor's recommendation, *Wallaces' Farmer* deemed it "absolutely sound national policy that agricultural products should be carried far cheaper than other products as long as they are selling so much lower in price." [7]

At first, Secretary Wallace attempted to gain voluntary acceptance of his preferential rate proposal by the railroads, but with understandably small success, and soon turned to government regulation as a possible means of compelling them to accept his ideas. On this matter, however, he had to move carefully, for the Agriculture Department had no official connection with the Interstate Commerce Commission, through which rate adjustment would have to be secured. Simply as a matter of propriety, Wallace did not want it to appear that he was assuming a position as an advisor to the commission, and he was careful to avoid addressing his comments directly to that agency. Still, he persistently advanced the proposal in an effort to gain support and

[6] Address before the Knife and Fork Club, Kansas City, Missouri, October 23, 1922, copy in Henry C. Wallace Papers, State University of Iowa Library, Iowa City, Box 3; and address at Davenport, Iowa, October 4, 1922, copy in Herbert C. Hoover Papers, Herbert Hoover Presidential Library, West Branch, Iowa, 1-I/2.

[7] *Wallaces' Farmer* 47 (November 10, 1922): 1325.

publicity for it and, presumably, to develop public pressure to force action.[8]

After the rate increases of 1920, there was sufficient protest from farm areas to induce the Interstate Commerce Commission to hold a series of hearings the next year for the purpose of reconsidering its earlier decision. Resulting in a limited acceptance of the preferential theory, the hearings were followed by orders for small reductions in freight charges on grain, vegetables, hay, and livestock, among other nonagricultural goods. Agitation continued, however, and in May 1922 the commission called for a 10 percent decrease in all rates not lowered the previous year.[9] Although Wallace was pleased with this action, he reminded those who looked upon the ruling as a final solution to the farmers' transportation problems that "rates still remain far above prewar levels and constitute a heavy burden on agriculture." [10] While any relief was welcome, furthermore, the commission's blanket reduction hardly conformed to the secretary's recommendation for preferential treatment of farm commodities. As agricultural conditions failed to improve and even declined in the western wheat regions, Wallace renewed his demands with increasing urgency. Finally, in September 1923, he pointedly enunciated his idea of preferential treatment by calling for a 25 per cent decrease in rates on all farm commodities. At a rather stormy cabinet meeting, the irritated secretary flatly asserted that the government was obligated to aid agriculture through its period of hardship and that this required, among other things, a re-

[8] Address before the Knife and Fork Club, Kansas City; "Report of the Secretary," USDA *Yearbook of Agriculture*, 1922, pp. 4–5; and Wallace, *The Wheat Situation*, pp. 23–24.

[9] Interstate Commerce Commission, *Annual Report*, 1921, p. 7; and Interstate Commerce Commission, *Annual Report*, 1922, pp. 18–19.

[10] "Report of the Secretary," USDA *Yearbook of Agriculture*, 1922, pp. 4–5.

duction by at least one quarter in existing freight charges.[11] Interestingly enough, the Interstate Commerce Commission was holding additional hearings on railroad rates at the time. Unmoved by Wallace's plea, the commission refused to order any further rate decreases.

Shortly after assuming the presidency, Calvin Coolidge demonstrated a passing interest in the preferential plan when he suggested to the railroads in October 1923 a voluntary cut in charges on farm products. But when the lines refused to accept the idea he showed no inclination to pursue the matter further. The President seemed more concerned about how such an arrangement would affect western roads, which carried mainly agricultural commodities, than he was with finding a solution to the farmers' transportation problems.[12] Partly due to Wallace's persistence, though, the proposal for preferential rates had gained support among agrarians and in Congress. As a result, a resolution was introduced into the Senate in 1924 and adopted the following year which declared that agriculture was a "basic industry" and directed the Interstate Commerce Commission to effect changes in rate schedules which would allow the transportation of farm products at the "lowest possible rates." [13] Even before Wallace's death in October 1924, however, interest in preferential rates was being replaced by agitation for more direct solutions to the problems of agriculture, and little progress was made toward implementation of the congressional resolution.

Wallace had a good deal more control over the regulation of

[11] "Notes of Cabinet Meeting, September 25, 1923"; and *New York Times,* September 26, 1923, pp. 1, 4.

[12] *New York Times,* October 31, 1923, p. 24 and November 2, 1923, p. 19; and Howard H. Quint and Robert H. Ferrell (editors), *The Talkative President* (Amherst: University of Massachusetts Press, 1964), p. 116.

[13] United States Congress, Senate, *Declaring Agriculture to Be Basic Industry of the Country,* Report 313, 68th Cong., 1st Sess. The measure was the Hoch-Smith Resolution.

commodity exchanges than of railroads, for the Agriculture Department had been charged with the administration of the grain futures trading acts. In his supervision of the exchanges, the secretary sought to apply his concept of regulated capitalism and the ideas of government-business cooperation which had been central to one strand of progressivism in the prewar period.

The conspiracy theory had explained away the farmers' troubles in earlier times of economic crisis, and in the 1920's it was once again called upon to make simple sense out of a complex situation which had left agriculture in a depressed condition. Low prices, according to grain producers, were the result of manipulation on the commodity exchanges, particularly in the trading in goods for future delivery. As irate farmers saw the situation, dealers contracted at current levels to supply grain at some future date and then proceeded to use their influence over the market to drive down prices so that they might purchase that grain at the lowest possible level. To accomplish this they might either circulate false rumors which would have a bearish effect on prices or suddenly dump large grain supplies under their control onto the market. With some truth and considerable appeal, the conspiracy theory gained a large bloc of adherents and resulted in widespread interest in federal regulation of futures trading. Responding to agrarian demands, Congress passed the Grain Futures Trading Act of 1921, providing for supervision of the industry by the Secretary of Agriculture and for a prohibitive tax on grain sold for future delivery by anyone other than the producer.

As already pointed out, Wallace was pleased with the act, had aided in revising the original bill, and had lent his support to help secure its passage. In administering the new law, he was quick to assure dealers that it was not the purpose of his department to prevent futures trading, hedging, and legitimate speculation or to interfere with the normal and proper operation of the exchanges, all of which functions he considered essential to farmers. "Our work," he promised, "will be of a constructive and thoroughly sympathetic sort." Specifically, the objective was "to make sure that the great public markets for grain . . . are free and open and that abuses and unfair charges are elminated." This required the elimination of "unfair price manipulation" and an end to the "circulation of unfounded rumors set afloat by designing operators." Emphasizing that "legitimate operators need have no fear of unwarranted meddling with their business," Wallace requested the cooperation of the exchanges in the administration of the law, pointing out to the dealers that the elimination of questionable practices in the grain market would be as much to their advantage as to the producers'. Commenting on such an arrangement, he expressed the belief that "surely the time has come when everyone must see that there must be a far larger degree of cooperation between the various large groups in this country which have interests in common. . . ."[14] Agriculture Department policy, therefore, was not to be based on the unilateral use of federal power against the trade but on a union of government and honest businessmen in opposition to unscrupulous manipulators.

Accepting Wallace's proposal for government-business cooperation, the Chicago Board of Trade, the country's largest and most influential grain exchange, gave its "assurance that the law would be given a fair test. . . ." But a minority within the board's membership, composed of small commission merchants dealing largely in grain consignments from farmers, local elevators, and county dealers, was determined to prevent enforcement of the act. Convinced that federal regulation threatened to de-

[14] Address before the Millers' National Federation, Chicago, June 29, 1921, copy in Edwin T. Meredith Papers, University of Iowa Library, Iowa City, Box 10; and address at Washington Court House, Ohio, October 18, 1922, copy in Gilbert N. Haugen Papers, State Historical Society of Iowa, Iowa City.

stroy their businesses, members of this group circulated a petition urging litigation to enjoin operation of the law and presented it to the officers of the Chicago board.[15] Failing to gain the support of the board, they proceeded on their own and in late 1921 brought suit against Wallace and the Agriculture Department. Rendered on May 15, 1922, the decision of the Supreme Court in *Hill* v. *Wallace* declared the law unconstitutional on the grounds that the levy on futures transactions was an unwarranted use of the taxing power. A short time later Congress passed the Grain Futures Act to replace the legislation rejected by the Supreme Court. Based upon the power to regulate interstate commerce, the new law also charged the Secretary of Agriculture with responsibility for supervision of the grain business, but eliminated the taxing provision. Equally satisfied with the second act, Wallace proceeded to apply the same interpretation to it as he had to the first.[16]

While the Agriculture Department's regulation of the grain trade failed measurably to improve the agricultural situation, it at least served to repudiate the conspiracy theory as applied to the operation of the exchanges. Manipulation proved to be much less widespread than excited grain farmers had imagined, and, as the few offenders were ferreted out, there was no noticeable effect on prices. Like other popular analyses of the agricultural crisis, the conspiracy theory merely clouded the issue and prevented farmers from seeing the real causes of their difficulties. Despite the fact that regulation did little to mitigate agrarian conditions, it helped to regularize and improve the efficiency of the country's grain trade and, particularly important from Wallace's viewpoint, to prevent adoption of the more radical proposal for federal ownership of the exchanges.

[15] Joseph P. Griffin to Wallace, October 29, 1921, Files of the Secretary of Agriculture, National Archives.
[16] *Des Moines Register,* November 1, 1922, p. 8.

Secretary Wallace most clearly demonstrated his concept of federal regulation in the administration of the Packers and Stockyards Act. Soon after Congress had enacted the measure, the Department of Agriculture established a special enforcement division. Applying the same ideas as in the case of the grain trade, Wallace instructed the new agency that "the purpose of the law was to be constructive; that we should try and bring [sic] about thorough cooperation between all of the agencies which have anything to do with producing and marketing live stock. . . ." While admitting that he felt a "responsibility to the packers and stockyards and the commission merchants who have built up large businesses," he indicated his intention "to deal fairly with all and to enforce the terms of the act without fear or favor." [17] Under Wallace's direction the policy was to be one of "self-policing," in which the Agriculture Department would aid packer and livestock concerns in regulating their own operations.[18] Although meat interests, to be sure, would have preferred no legislation at all, Wallace was "agreeably surprised . . . [with] the spirit in which they took the passage of the packer bill" and confident that they would "accept it in good faith and observe its provisions." [19] Such an attitude, of course, was essential to the government-business cooperation which the secretary envisioned.

The Packers and Stockyards Administration focused initially on the activities of commission merchants. Representing producers at central stockyards, these agents performed services upon which farmers not in a position to accompany livestock to market were highly dependent. Why the Agriculture Department chose to concentrate first on this group rather than the meat in-

[17] Wallace to Charles E. Collins, May 12, 1923, Files of the Secretary of Agriculture.
[18] "Minutes of the Meeting of Bureau Chiefs, October 3, 1921," *ibid;* and address at Washington Court House, Ohio.
[19] Wallace to Harding, August 10, 1921, Files of the Secretary of Agriculture.

dustry itself is uncertain, but it might have been because the commission men were considered more vulnerable than the powerful packers. At any rate, the department could justify its action by pointing out that these agents were, for the most part, under the control of the big packing companies. Wallace urged the commission merchants at the major centers to formulate uniform rate schedules for themselves, which his office could approve and enforce. But department officials negotiating with agencies at several of the large stockyards found them unsympathetic to standardized charges.[20] Also investigated by the Packers and Stockyards Administration were reports of malpractice among commission men at several livestock centers, which resulted in charges brought against a number of groups for discrimination in boycotting marketing associations and for other irregular practices.[21]

By and large, the department's action in regard to the commission agents was disappointing. With the merchants disinclined to accept Wallace's policy of self-policing and cooperation, the work of the Packers and Stockyards Administration was complicated considerably. Due to the diversity of conditions throughout the yards, the establishment of uniform rates and the prevention of unethical practices were difficult, if not impossible, tasks for a federal agency to accomplish on its own.[22] Significantly, it was a firm of commission merchants and livestock dealers, rather than meat packers, which brought suit in 1922 to test the Packers and Stockyards Act, a case which resulted in a decision upholding the constitutionality of the law. Not entirely popular among

[20] "Minutes of the Meeting of Bureau Chiefs, October 3, 1921," *ibid*.
[21] Wallace to Sydney Anderson, March 12, 1922; and "Minutes of the Meeting of Bureau Chiefs, April 3, 1922," *ibid*.
[22] "Minutes of the Meeting of Bureau Chiefs, October 3, 1921"; Chester Morrill to W. A. Jump, January 31, 1922, *ibid.;* and United States Congress, House, Committee on Expenditures in the Department of Agriculture, *Hearings on the Packers and Stockyards Act,* 68th Cong., 1st Sess., pp. 182–83.

farmers, the department's preoccupation with the commission agents led one critic to observe, after the act had been in force for nearly a year, that "the activities of the Department of Agriculture seem to be revolving around the commission men, whereas the packer was supposed to be the wicked one in the livestock commerce." [23]

The Packers and Stockyards Administration also concentrated a good deal of attention on the question of the examination of books and records. Of the opinion that effective enforcement of the law required ready access to all pertinent company information, the Agriculture Department first of all investigated the accounting systems of the major packers to determine if they disclosed the necessary data. Although some farmers were demanding that the meat industry be required to install uniform bookkeeping methods, Wallace felt that this would be unfair if the systems in use proved adequate to the needs of his department.[24] Apparently the investigation satisfied him, for no action was taken to impose standardized accounting. Next, the Department of Agriculture proposed permanently to assign an accountant of its own to each of the large packers, the work of which would be to make detailed examinations of all accounts, records, and memoranda. Upon the instruction of Wallace's office, he was to review and verify financial books, determine ownership and control, and ascertain the nature and extent of connections with other packers and related concerns. In addition, he was to analyze accounting methods and practices and to become familiar with all managerial details.[25] There would be little, in short, that the Agriculture Department's accountants would not know about the operations of their respective companies.

[23] J. Ralph Pickell to Wallace, August 28, 1922, Files of the Secretary of Agriculture.
[24] Wallace to Benjamin C. Marsh, November 23, 1922, ibid.
[25] H. M. Bain to Charles E. Herrick, March 29, 1923, ibid.

Although most of the meat packers complied with the department's request, two of the Big Five, Swift and Cudahy, refused to open their books. The act was vague on the authority of the department to demand such complete access to records, and the two companies argued that the Packers and Stockyards Administration was not empowered to investigate books unless there was evidence the law had been violated. Determined to resist, they ignored Wallace's persistent demands and his threats to bring court action if they failed to comply with the department's policy.[26] Insisting that an audit was not only legal but necessary for his office to discharge its duty, the secretary took the position that, if the packers had nothing to hide, they should have no objections to opening their records. In December 1923 he told a meeting of the American Farm Bureau Federation that "my experience of the past year and a half has shown me very clearly the imperative need of knowing the packers' books." Early the following year Wallace requested Attorney General Harry M. Daugherty to initiate a suit against the two companies. "You may be sure," Wallace wrote to a friend, "that I am not going to leave anything undone to establish our right to examine the packers' books." [27] The companies involved were able to delay litigation until after Wallace's death, however, and the case was eventually dropped.

Wallace's most dramatic confrontation with the meat industry over enforcement of the Packers and Stockyards Act came as a result of the absorption of one member of the Big Five by another. In November 1922 J. Ogden Armour, president of the Armour Meat Packing Company, submitted a plan to Daugherty and Wallace for the purchase by his firm of the physical assets of

[26] Wallace to Edward L. Burke, December 6, 1923; USDA press release, December 11, 1923, copy in *ibid.;* and *Hearings on the Packers and Stockyards Act,* p. 133.

[27] *Oklahoma Live Stock News,* December 13, 1923, p. 1; and Wallace to A. H. Hartman, February 12, 1924, Files of the Secretary of Agriculture.

Morris and Company, and requested an opinion as to whether the proposed action would constitute a violation of either the antitrust or packers laws.[28] Though Armour might have been attempting to follow the policy of government-business cooperation set down by Wallace, it is more likely that he was only trying to avoid a move which would provoke further federal interference in the meat industry.

Writing to Harding a few days after the conference with Armour, Wallace indicated that, while he had not had time to consider the matter fully, "from the study already given it I have reached a fairly definite conclusion in my own mind"; he failed to say, though, what that conclusion was.[29] Wallace, Harding, and Daugherty met a short time later to discuss Armour's inquiry and decided that "there was no obligation . . . [upon the government] to endorse or acquiesce in the action proposed or to express an opinion concerning it." Harding directed the Attorney General, however, to study the question further and to write a formal opinion. In a published letter to Wallace, Daugherty announced his ruling that under the Packers and Stockyards Act the Secretary of Agriculture was not required to advise the meat industry "in regard to such a transaction in advance of its consummation." The act "contemplates action only when you [Wallace] have reason to believe that any packer has violated or is violating the provisions of that act. . . ."[30]

On December 12, 1923, Wallace issued a rather vague statement to the press giving his decision on the Armour proposal. Referring to Daugherty's opinion, he declared that there was apparently no occasion for action under the Packers and Stockyards Act "at the present time," but went on to state that he had

[28] *New York Times,* November 16, 1922, p. 1.
[29] Wallace to Harding, November 18, 1922, Warren G. Harding Papers, State Historical Society of Ohio, Columbus, Box 1.
[30] Wallace to President of the Senate, December 12, 1922; and Daugherty to Wallace, December 9, 1922, Files of the Secretary of Agriculture.

found nothing in the law prohibiting Armour from acquiring the physical assets of Morris and Company. While the "probable effect" on competition of the proposed purchase suggested no need for immediate action, Wallace admitted that there was some question as to whether it would bring about conditions in violation of the law. He indicated that the Agriculture Department was studying that aspect of the matter. Despite the widely held belief to the contrary, the secretary explained, Armour had not made an application for the merger of the two companies, but rather had asked the department's attitude on his purchase of the physical assets of Morris in order to expand his own operations.[31] The distinction, however, would be lost on many who could not see the practical difference between the two situations; in either case the Morris company would disappear and its competition would be eliminated.

That Wallace's stand on the Armour issue lacked something in the way of clarity was attested to by the large volume of conflicting correspondence his department received on the subject. An Illinois banker was "well pleased" with the Agriculture Department's sympathetic attitude toward the merger and was of the opinion that "our Government should aid and encourage business enterprises; and not so act as to hamper and discourage them. . . ." A National Chamber of Commerce officer praised Wallace's position "that the government will not interfere with business so long as business is honestly and properly conducted . . . [as] a great step forward," while the American Wholesale Grocers' Association wrote that it was "especially pleased with what appears . . . as your very logical decision" in sanctioning the proposed merger. A. Sykes of the Corn Belt Meat Producers' Association, on the other hand, was "glad to note the press despatches [sic] from Washington say that you [Wallace] are not fa-

[31] USDA press release, December 12, 1922, copy in *ibid.;* and *New York Times,* December 13, 1922, p. 16.

vorable to the Packer merger. . . ." Still another correspondent wanted to express his "appreciation for the heroic stand you are taking against the merger of Armour and Morris and Co." [32]

Armour chose to interpret the secretary's statement of December 12 as assurance that the government would not interfere with his acquisition of the Morris Company. In late December 1922 he created a new corporation, Armour and Company of Delaware, and, although the statement of incorporation made no mention of the contemplated merger, it soon became clear that the new organization was to act as a holding company in the forthcoming transaction. Early the following year Wallace learned of the action and was told that shortly Armour would absorb Morris and Company.[33]

News of the approaching merger resulted in a wave of protest from livestock producers. They concluded that the Agriculture Department had betrayed them by taking no action to head off the absorption of Morris and Company and thus to prevent the resultant lessening of competition in the meat industry. Senator George W. Norris of Nebraska introduced a resolution ordering the Federal Trade Commission to investigate the matter and report its findings to Congress. At the hearings on the resolution in February 1923, Wallace carefully explained once more that Armour's original inquiry concerned only a proposal for the purchase of Morris' physical assets, but he again failed to clarify why this action was less objectionable than an outright merger. It was not until later, the secretary told the Senate agricultural committee, that he had learned of the plan for joining the two companies, after which the Agriculture Department had begun

[32] William George to Wallace, December 16, 1922; John H. Camlin to Wallace, December 14, 1922; J. H. McLauren to Wallace, December 13, 1922; A. Sykes to Wallace, December 1, 1922; and S. T. Stone to Wallace, December 11, 1922, Files of the Secretary of Agriculture.

[33] *Hearings on the Packers and Stockyards Act*, p. 130.

proceedings to prevent the transaction. Contending that intervention by the Federal Trade Commission would hinder the action already initiated by the Packers and Stockyards Administration, Wallace urged that the Norris resolution be discarded.[34] In compliance with his request, the committee laid the resolution aside.

The action contemplated by the Department of Agriculture was revealed on February 26, 1923, when Wallace issued a notice of complaint advising Armour and Morris that, in his opinion, their proposed merger would have "the tendency or effect of restraining interstate commerce or of creating a monopoly. . . ." He judged that the result would be materially to increase the power of Armour "to control and dominate the live stock and meat packing industry. . . ." As prescribed under the Packers and Stockyards Act, therefore, the Secretary set April 2 for the beginning of hearings to determine if he should issue a ccase and desist order in the case.[35] Accompanying the notice of complaint was a press release relating the sequence of events leading up to the department's formal action. In this statement Wallace claimed that two days after his November conference with Harding and Daugherty he had advised Armour against taking over the Morris assets. "I told him that our investigation had not been completed," the secretary reported in the press release, "but from results so far my decision probably would be unfavorable to the proposed purchase." Wallace remembered that on two other occasions he had repeated his warning against the transaction, once

[34] United States Congress, Senate, Subcommittee of the Committee on Agriculture and Forestry, *Hearings on a Resolution Directing the Federal Trade Commission to Investigate and Report to the Senate the Facts as to the Proposed Organization and Merger of Corporations of Armour and Company*, 67th Cong., 4th Sess., pp. 2–6; see also Wallace to Edward Voigt, February 23, 1923, Files of the Secretary of Agriculture.

[35] "Notice of complaint in the matter of violations of Title II of the Packers and Stockyards Act, 1921," copy in files of the Secretary of Agriculture; and *Des Moines Register*, February 23, 1923, p. 7.

in early December and again on December 11.[36] Despite the fact that Wallace had ample opportunity and reason to reveal these efforts to persuade Armour to abandon his plan, this was the first mention of any of them. Why he chose to withhold such information from his statement of December 12 or from the committee hearings on the Norris resolution remained unanswered. This belated reference to the warnings to Armour, if indeed there had been any warnings at all, called into question the validity of the secretary's contentions and sincerity of his stand. Wallace's plans to investigate notwithstanding, Armour and Morris carried through their merger and announced its completion on March 28.[37]

After an extension granted to the packers to prepare their case, hearings before the Packers and Stockyards Administration began on April 30 in Kansas City, Missouri. Other hearings were subsequently held in East St. Louis, Omaha, Chicago, and Washington. During the protracted series, lasting for one and a half years, testimony was taken from representatives of producers, packers, stockyards, commission men, and others. Upon completion of the hearings, the case was to be argued before Wallace, who would then decide on the basis of the evidence gathered if the law had been violated and whether or not a cease and desist order should be issued.[38] The secretary died before the final phase of the investigation, however, and the decision was left to his successor, William M. Jardine, who brought the matter to a

[36] "Statement of Secretary Wallace Concerning Complaint issued Against Armour and Company with Reference to their Acquisition of the Properties of Morris and Company," USDA press release, February 26, 1923, copy in files of the Secretary of Agriculture; and *New York Times,* February 27, 1923, p. 8.

[37] Telegram, Chester Morrill to Wallace, March 29, 1923, Files of the Secretary of Agriculture; and *Washington Post,* March 29, 1923, p. 5.

[38] Telegram, Chester Morrill to Wallace, March 22, 1923; Wallace to A. Sykes, May 12, 1923; and Wallace to Ray Anderson, February 4, 1924, Files of the Secretary of Agriculture.

close on September 14, 1925, by dismissing the complaint against the merger. According to the new secretary, Armour's acquisition of Morris did not "in itself constitute a violation of the law . . . ," for there had been "no apparent lessening of competition since the merger." [39]

No other aspect of Wallace's tenure as Secretary of Agriculture drew so much criticism as the administration of the Packers and Stockyards Act. Producers who had applauded his opposition to the meat industry as an editor and as secretary of the Corn Belt Meat Producers' Association now bitterly attacked his enforcement of the law. One disillusioned farmer was at a loss to explain why a man who had been "the champion of the cause of agriculture" was now "functioning in a way highly satisfactory to the packing industry." [40] As old and trusted a friend as A. Sykes had to tell Wallace that his administration of the packers law had caused "resentment and loss of confidence . . . among producers." The secretary's handling of the meat industry, declared another representative of agriculture, demonstrated that he was "one of the worst enemies of the farmers in the United States." A Minnesota congressman reported that Wallace had become most unpopular among the farmers of his district because of the belief that "he has favored the interests, the middle-man, at the expense of the producer, and that that belief is well founded." [41]

While there was some complaint concerning the handling of commission men and the failure to compel packers to install uniform accounting systems, the major part of the criticism arose in response to the Armour-Morris merger. For the most part, live-

[39] "Conclusion and Order," September 14, 1925, copy in *ibid.*
[40] *Hearings on the Packers and Stockyards Act*, pp. 25–26.
[41] A. Sykes to Wallace, June 18, 1923, Files of the Secretary of Agriculture; United States Congress, Senate, Committee on Agriculture and Forestry, *Hearings on the Purchase and Sale of Farm Products,* 68th Cong., 1st Sess., p. 323; and Halvor Steenerson to C. Bascomb [sic] Slemp, August 3, 1923, Calvin Coolidge Papers, Manuscript Division, Library of Congress, File 1, Box 1.

stock producers were of the opinion that Wallace had the power under the Packers and Stockyards Act to prevent the transaction. A group calling itself the Peoples' Reconstruction League expressed the feelings of many farmers when it attacked the secretary for failure to use "his unquestioned power . . . to obtain an order restraining the merger." [42] An official of the Packers and Stockyards Administration, who resigned as a result of his disagreement with its policy, told a House committee that the Agriculture Department had "failed to do what I think could have been done at the right time to prevent the merger of Armour and Co." Although he doubted that the Packers and Stockyards Act provided the legal means to block the transaction, he regarded it "primarily as a publicity statute," under which "the Secretary had a beautiful opportunity . . . to have told the country that that sort of combination was in violation of the [law]. . . ." Presumably, Wallace could have generated sufficient public opposition to the proposed merger to have discouraged Armour and Morris from carrying it through.[43] Benjamin C. Marsh, managing director of the militant Farmers' National Council, charged that the secretary had initiated proceedings against the packers only after pressure was applied by livestock producers, but by then it was too late to prevent the merger. And even those who agreed that the department had moved as soon as legally possible were concerned over "the long delay in bringing the case to a termination. . . ." [44]

Much of the criticism revolved around two of Wallace's appointments. Chester Morrill, head of the Packers and Stockyards Administration, and Charles J. Brand, consulting specialist in marketing, were identified by the Farmers' National Council as

[42] Peoples' Reconstruction League press release, June 4, 1923, copy in Harding Papers, Box 1.
[43] *Hearings on the Packers and Stockyards Act*, pp. 19–21.
[44] Benjamin C. Marsh to Wallace, October 6, 1924; and O. M. Kile to Wallace, May 1, 1924, Files of the Secretary of Agriculture.

two of the meat packers' "warmest friends." [45] The charge against Morrill seemed to be unfounded. Since entering the Agriculture Department in 1914, he had held several posts but had never been in close contact with the meat industry. Brand, on the other hand, had been chief of the Bureau of Markets from 1913 to 1919, and his alleged implication in the efforts of the packers to prevent the Federal Trade Commission's investigation of the meat industry after the war had aroused the suspicion of farmers. He left the department in 1919 and returned at Wallace's request in 1922. When rumors of Brand's imminent reappointment began to circulate, at least two prominent congressional farm representatives, Nebraska's Senator George W. Norris and Representative J. N. Tincher of Kansas, advised Wallace against it.[46] Although Brand was not "a well-known tool of the packers," as Benjamin C. Marsh claimed, he was sympathetic toward them, which of course opened to question his appointment as consulting specialist in the regulation of the meat industry.[47]

With mounting criticism of the Packers and Stockyards Administration, the House Committee on Expenditures in the Agriculture Department held hearings on the matter in April and May 1924. The principal witness was John M. Burns, a law clerk in Wallace's office. In a lengthy testimony, Burns charged that the packers law was being enforced in a way highly favorable to the meat industry and that the Agriculture Department was stifling the competition which the act was designed to foster. In an obvious reference to Morrill and Brand, he attributed the situation to the influence over Wallace of certain department of-

[45] Farmers' National Council press release, August 30, 1924, copy in *ibid.*

[46] George W. Norris to Wallace, July 14, 1922; and J. N. Tincher to Wallace, January 20, 1923, *ibid.*

[47] *Hearings on the Purchase and Sale of Farm Products,* p. 320. For Brand's reply to charges see Brand to Edward J. King, June 7, 1924, Files of the Secretary of Agriculture and reprinted in *Hearings on the Packers and Stockyards Act,* pp. 193–98.

ficials who were favorable toward the meat interests.[48] Burns was dismissed a short time later, purportedly as part of a personnel reduction program to cut operating costs, and critics interpreted this action as further evidence of the influence of the packers over department policy.[49]

Wallace revealed an ineptness in adminstrating the Packers and Stockyards Act that was not characteristic of his secretaryship in general. Much of the trouble, it would seem, arose from limitations in the law itself, limitations which prevented the kind of regulated capitalism preferred by the secretary. He admitted as much in the statement accompanying his complaint against the Armour-Morris merger in February 1922. If it were within the department's power to insure that only a fair profit would be taken, he explained, "it is conceivable that such a combination as proposed could act to the benefit of the producer and consumer." [50]

As noted, Wallace preferred the stronger Kenyon-Kendrick bill to the packer control measure eventually enacted, primarily because its licensing provision would have given him the authority necessary to supervise combinations of the sort created by the union of Armour and Morris. The real problem in regulation stemmed from a conflict of objectives between the law and those responsible for its administration. While the Packers and Stockyards Act was designed, as John M. Burns correctly testified, to foster competition in the meat industry, Wallace and

[48] *Hearings on the Packers and Stockyards Act,* pp. 1–15, 39–51, 60–72.

[49] "Statement of the Solution of the Department of Agriculture Relative to the Separation of Mr. John M. Burns from the Service of the Department," ms.; and Benjamin C. Marsh to Coolidge, August 20, 1924, Files of the Secretary of Agriculture. Burns continued his public attacks after dismissal from the department. See Wallace to Charles E. Hearst, September 11, 1924, Charles E. Hearst Papers, University of Northern Iowa Library, Cedar Falls.

[50] "Statement of Secretary Wallace Concerning Complaint Issued against Armour and Company with Reference to their Acquisition of the Properties of Morris and Company."

top officials of the enforcement agency were not concerned about competition. They preferred instead a system permitting consolidation under strict federal regulation, the basis of which was to be a permanent government-business cooperation, with government of necessity being the dominant partner. But the Packers and Stockyards Act provided for temporary federal intervention to restore competition and then for withdrawal from the economy; government was to be a watchdog not a partner.[51]

In Wallace's effort to impose his ideas upon the legal framework of the Packers and Stockyards Act, he not only failed to develop the system of regulation he desired, but hampered the enforcement of the law as well. The secretary attempted to create a condition of government-business cooperation without having the federal power upon which that cooperation had to rest. For a time Wallace thought, or perhaps only hoped, that there was a possibility of accomplishing his purpose. "We have got along very well," he told the American Meat Packers' Institute in September 1923. "I want to express my appreciation . . . for the manner in which you have met our people. . . . The relations have been fine." [52] But his trouble with the commission merchants and his failure to gain access to the records of all packers testified to the fact that the voluntary cooperation of the meat interests was hardly satisfactory and demonstrated the department's weakness under the law to force compliance.

The conflict over the Armour-Morris question illustrated further the differences between Wallace's idea of regulation and the objectives of the packers law. There is little doubt that the secre-

[51] Morrill early realized the inadequacies of the law and warned soon after its enactment that "it would be impossible to show spectacular results in the administration of the act very soon, and perhaps never." "Minutes of the Meeting of Bureau Chiefs, October 3, 1921"; Morrill to W. A. Jump, January 31, 1922, *ibid.*

[52] Address before the American Meat Packers' Institute, Atlantic City, September 22, 1923, copy in *ibid.*

tary favored the merger and that he gave his tacit approval, or at least withheld final judgment, when first approached on the matter in November and December 1922. Apparently hoping that either through the control law or voluntary cooperation on the part of the packers the resulting combination could be adequately regulated, he was reluctant to discourage a consolidation which he felt offered the possibility of benefits to both producers and consumers. Despite his subsequent statements to the contrary, it is likely that he did not arrive at his ultimate opposition until February 1923. No doubt his change of position was due in part to mounting criticism among producers. More important in Wallace's decision to initiate proceedings against the merger, however, were probably a growing awareness of the inadaptability of the Packers and Stockyards Act to his theory of regulation and an increasing uneasiness over the packers' unwillingness to cooperate in their own supervision.[53] In any case, Wallace's failure to make definite his opposition at the outset obstructed enforcement of the law, of which the merger was a clear violation. Earlier action on the secretary's part might not have prevented the union of Armour and Morris, but it would have established the department's position from the beginning and made more effective its eventual move against the merger.

The disappointing results of the Packers and Stockyards Act, therefore, were not due to Wallace's capitulation to the packers, as some critics thought, nor were they a payoff to the meat industry for campaign donations in 1920, as others charged.[54] They were largely the outgrowth of an abortive effort to apply a theory

[53] Donald R. Murphy disagrees with this interpretation. He maintains that Wallace would never have favored the merger and that criticism among livestock producers resulted from a misinterpretation of his position. Interview with Donald R. Murphy, Des Moines, Iowa, August 29, 1965.

[54] Benjamin C. Marsh to Wallace, October 6, 1924, Files of the Secretary of Agriculture; and *Hearings on the Purchase and Sale of Farm Products,* p. 187.

of regulation without the legal machinery necessary to carry it through.

On the matter of business regulation, Secretary Wallace found himself in greater disagreement with the farmers he was representing than on any other question. Traditional resentment of big business had led the large majority of agrarians to endorse programs for the dissolution, rather than regulation, of trusts. Though his aversion to corporate wealth was no less intense, Wallace nevertheless felt that the trust-busting solution to the problems of maturing industrialism was inappropriate and self-defeating. First of all, he held, business consolidation was an integral part of an advanced capitalistic economy, and thus attempts to check the movement in the United States were bound to fail. Secondly, he believed that if the government dissipated its energies on trust-busting, it would miss a very real opportunity to regulate corporate wealth for the common good. Agrarians generally failed to understand, much less accept, this analysis of the American economy, and their feeling that Wallace had abandoned them and gone over to the interests is not difficult to understand.[55]

Just as the agricultural economics program, improved rural credit facilities, and higher agricultural tariffs failed measurably to improve the lot of the farmers, so federal regulation fell short of Wallace's expectations. Transportation costs, it was true, were moderately reduced, and some questionable practices among grain dealers, commission merchants, and meat packers had been eliminated. But the larger benefits to the country generally and to

[55] Not all producers opposed the merger. James R. Howard, president of the Farm Bureau, thought that "it may be a good thing. Since the control bill is in operation there is no danger to the public from monopoly which existed a few years ago and consolidation certainly would have the effect of cutting down a lot of expensive overhead in the packers' distribution of their product." J. R. Howard to Wallace, December 15, 1922, Files of the Secretary of Agriculture.

farmers in particular which Wallace assumed would flow from regulation did not materialize. Part of the problem, to be sure, lay in inadequate legislation and inexperienced bureaucratic administration. But the agricultural crisis stemmed from something more fundamental than the abuses of middlemen and, as Wallace would soon have to admit, could not be solved, nor for that matter mitigated, through federal regulation.

Chapter X / ANOTHER
FIGHT WITH HOOVER

Wiith the increased dependence of American farmers on export markets in the postwar period, it was to be expected that the Department of Agriculture would focus considerable attention on foreign production and demand as part of its agricultural economics program. If producers were to make the necessary adjustments in output, Wallace reasoned, they would have to be kept well informed on conditions abroad. Thus the collection, interpretation, and dissemination of data on agricultural production and demand throughout the world became an important part of the work of the BAE in the early 1920's. At the same time Secretary of Commerce Herbert C. Hoover was beginning a similar program in his own department. Accepting appointment to Harding's cabinet on the condition that he be permitted to reorganize the Commerce Department and bring under its control all functions which he felt properly belonged to it, Hoover had made quite clear that this was to include a "free hand to concern myself with the commercial interests of farmers. . . ." [1] Cast against the existing antagonism between Wallace and Hoover stemming from their conflict during the war, competition for ju-

[1] Telegram, Hoover to Harding, February 24, 1921, Warren G. Harding Papers, State Historical Society of Ohio, Columbus, Box 698; see also Herbert C. Hoover, *The Memoirs of Herbert Hoover* (New York: Macmillan, 1952), vol. 2, p. 109.

risdiction over agricultural marketing was certain to cause serious interdepartmental friction.

For some time before Wallace assumed office the Agriculture Department had been engaged in foreign marketing work. Beginning in 1903 the department had a special European representative who collected and transmitted to the United States data on production and demand in farm commodities. With formation in 1910 of the International Institute of Agriculture in Rome, which was primarily a reporting agency, this post was temporarily discontinued.[2] The act of 1913 creating the Office of Markets specifically authorized the Secretary of Agriculture to acquire and disseminate information on the marketing and distribution of farm products, and it initiated the establishment of permanent departmental representatives in the foreign field to investigate supply and demand and to promote the export trade in American agricultural goods.[3] Due to inadequate appropriations and the disruption of the war, however, the program made little headway, and by 1921 the Agriculture Department had only two agents abroad, one in Europe and the other in South America. Most of the department's foreign agricultural information, therefore, had to come from outside sources, including official reports of foreign governments, the International Institute, and American consular officers and commercial attachés.[4]

Shortly after Wallace moved into the secretaryship, several officials of the department advised him that the existing reporting

[2] "Foreign Agricultural Information," ms. in files of the Bureau of Agricultural Economics, National Archives and Henry C. Taylor Papers, State Historical Society of Wisconsin, Madison.

[3] Unsigned memorandum to Project Leaders, January 9, 1919, BAE Files.

[4] George Livingston to Wallace, March 12, 1921; George Livingston to Wallace, March 23, 1921; Leon M. Estabrook to Wallace, March 14, 1921; Frank Andrews to Leon M. Estabrook, March 12, 1921, *ibid.;* and "Outstanding Features of the Foreign Work of the Bureau of Agricultural Economics," ms. in files of the Secretary of Agriculture, National Archives.

system was inadequate to meet the needs of the country's farmers. Foreign government reports were often insufficient, of questionable accuracy, and released too late to provide a usable source of information. Though serving a vital function, the International Institute failed to supply enough data to give American producers a comprehensive picture of world conditions. Lacking technical training in agriculture, consular officials and commercial attachés could not be depended upon to furnish the kind of timely and specific information necessary to guide production adjustment. It was therefore imperative, recommended one advisor, that the department place a number of well-trained representatives in key positions abroad to report on conditions in the foreign market.[5]

Wallace had been thinking along similar lines for some time before entering Harding's cabinet, and this advice not only made economic sense to him but also appealed to his faith in expertise.[6] Toward the end of the preceding Congress, the department had been granted a special appropriation of $50,000 for collecting and disseminating information relative to world agricultural conditions, and Wallace decided to use this to begin the expansion of foreign information services. "While I think that we should not undertake to execute any elaborate plans . . . ," he wrote to Henry C. Taylor, "I do believe that we should at the beginning block out the project as completely as we can." Consistent with his idea of government-business cooperation, the secretary suggested that it would be well to secure the advice of other interests concerned with foreign agricultural markets, such as the packers, grain dealers, and cotton exporters.[7]

[5] Leon M. Estabrook to Wallace, March 14, 1921; George Livingston to Wallace, April 22, 1921; and E. G. Montgomery to George Livingston, March 15, 1921, BAE Files.

[6] *Wallaces' Farmer* 43 (December 13, 1918): 1812.

[7] George Livingston to Wallace, March 12, 1921, BAE Files; and Wallace to Taylor, July 9, 1921, Files of the Secretary of Agriculture.

Taylor set to work immediately developing the program, and before the end of the year several projects were underway. The department sent E. C. Squire, a specialist in livestock marketing, to Europe to report on meat production and demand, and to investigate the possibilities of increasing the export trade of the United States. Working closely with the State Department, Squire secured a relaxation of restrictions on the importation of American pork into Great Britain, Germany, and the Netherlands and kept the Agriculture Department informed on the situation in the European meat industry.[8] After the close of the World Cotton Conference in 1921, part of the delegation from the Agriculture Department visited the major fiber importers on the Continent to study and report on the cotton market.[9] George F. Warren, a farm economist at Cornell University, and W. L. Callander, an official of the department, traveled throughout Europe in September and October 1921 studying conditions relative to the agricultural trade, arranging for prompt transmission of all available crop and livestock information to the Agriculture Department, investigating ways to facilitate the export of American farm products, and laying the basis for future foreign work of the BAE. Upon their return Warren and Callander recommended that the department place at least five permanent representatives at various points in Europe to gather and transmit data.[10] Louis G. Michael, a consulting specialist in marketing,

[8] E. C. Squire to Taylor, August 15, 1921, *ibid.;* memorandum, O. C. Stine to Taylor, undated; O. C. Stine to Taylor, January 25, 1923, Taylor Papers; "Outstanding Features of the Foreign Work of the Bureau of Agricultural Economics"; and "Report of the Secretary," USDA *Yearbook of Agriculture,* 1924, p. 49.

[9] Wallace to Hoover, May 24, 1921, BAE Files; and Wallace to Hoover, December 9, 1921, Files of the Bureau of Foreign and Domestic Commerce, National Archives, File 261.1, Box 1453.

[10] Memorandum, G. F. Warren to Taylor, September 9, 1921, BAE Files and Taylor Papers; and W. F. Callander and G. F. Warren to Taylor, October 30, 1921, Files of the Secretary of Agriculture.

spent two years in eastern Europe studying grain conditions to ascertain the probable effect on American farmers as this area returned to full production after the war.[11] Later, in 1923, Charles J. Brand toured Europe for the department on a mission similar to the Warren-Callander trip and, as had the earlier representatives, recommended stationing agricultural agents throughout the Continent, as well as in Asia and Australia.[12] In addition, the two permanent foreign representatives established earlier in Europe and South America continued to send regular reports on farm production in their areas.[13]

Despite the department's strong interest in establishing a network of agricultural agents abroad, however, there was little progress made. The main factor hindering further development of foreign work was a lack of funds, as appropriations for informational services abroad provided for special missions but were inadequate for the kind of permanent system recommended by Warren, Callander, and Brand. And efforts to gain larger grants proved futile, largely because of the feeling in Congress that expansion of the Agriculture Department's foreign services would duplicate work already carried on by the Commerce Department.[14]

While Taylor and the BAE moved to broaden overseas facilities, Hoover and Julius Klein, head of the Bureau of Foreign and

[11] Wallace to Charles Evans Hughes, September 15, 1921; Taylor to Wallace, January 22, 1923; and Louis G. Michael to Taylor, undated, Taylor Papers.

[12] "A Foreign Agricultural Service for the United States Department of Agriculture" (Brand's report), Files of the Secretary of Agriculture.

[13] Taylor to Julius Klein, September 13, 1921; Lloyd Tenny to Wallace, November 26, 1921, BAE Files; Edward F. Feely to Julius Klein, January 22, 1922, Files of the Bureau of Foreign and Domestic Commerce, File 155, Box 926; and Edward A. Foley to Taylor, March 31, 1923, Taylor Papers.

[14] Julius Klein to Edward F. Feely, November 30, 1921, Files of the Secretary of Commerce, National Archives, File 80553, Box 485; and Wallace to Taylor, September 1, 1923, Files of the Secretary of Agriculture.

Domestic Commerce, were instituting a parallel plan in the Commerce Department. Under a $250,000 appropriation for the improvement of its foreign field service, this department also chose to augment its activities in the collection and dissemination of agricultural information.[15] In May 1921 Hoover established farm commodity agencies under the Bureau of Foreign and Domestic Commerce, and three months later he created the Foodstuffs Division for the purpose of coordinating foreign survey and reporting work in farm products.[16] Alfred P. Dennis, a Commerce Department representative, traveled throughout Europe for three years studying conditions as they affected the export market for American agricultural goods and made an exhaustive report upon his return in late 1923.[17] At the same time commercial attachés increased their activities in gathering and transmitting data on farm commodities.[18]

The activities of the Agriculture and Commerce Departments, of course, resulted in considerable duplication of work. Initially, the two agencies attempted to coordinate their functions, and in some cases with a measure of success. Upon Wallace's suggestion an interdepartmental conference met in July 1921 for the purpose of formulating a basis of cooperation and a plan for avoiding duplication. At the meeting Klein was adamant in his insistence that commercial attachés were well prepared to gather and

[15] Taylor to Wallace, July 29, 1921, Files of the Secretary of Agriculture.

[16] "Report on Encroachments of Commerce on Agriculture," ms. in Taylor Papers; and *Annual Report of the Bureau of Foreign and Domestic Commerce,* 1922, p. 46.

[17] Alfred P. Dennis to Julius Klein, December 15, 1921, Files of the Bureau of Foreign and Domestic Commerce, File 155, Box 926; and Alfred P. Dennis to Coolidge, November 9, 1923, Calvin Coolidge Papers, Manuscript Division, Library of Congress, File 227D, Box 157.

[18] Edward F. Feely to Julius Klein, December 5, 1921, Files of the Bureau of Foreign and Domestic Commerce, File 371, Box 1867; C. F. Herring to Julius Klein, October 10, 1921, *ibid.,* File 155, Box 926; and Julius Klein to Leon M. Estabrook, December 15, 1921, BAE Files.

report agricultural information, and he rejected the contention that there was a need for permanent Agriculture Department representatives abroad. The bureau chief suggested that agricultural agents might be better utilized if they traveled from country to country advising commercial attachés on the type of data desired and investigating technical matters relating to agriculture. Reporting to Hoover on the conference, Klein indicated that "the chief problem was to persuade them [representatives of the Agriculture Department] not to establish field agents abroad and after some discussion they saw the advisability of not making such appointments." [19] Although Wallace's department was willing to allow Commerce representatives to handle trade promotion in food and fiber, the secretary held to the opinion that those agents were not trained to collect information on production and demand in farm commodities and, despite Klein's impression, that special foreign agricultural representatives were thus necessary.[20] But, due to insufficient funds for a complete program of its own, the Agriculture Department found it necessary to instruct its commissioners abroad to work through the commercial attachés, an arrangement with which Klein pronounced himself entirely satisfied.[21] For a while, at least, representatives of the two agencies collaborated on compiling reports, commercial attachés aided special foreign agricultural commissions, and information flowed freely between departments. The formation

[19] Wallace to Hoover, July 14, 1921, BAE Files; Taylor to Wallace, July 29, 1921, Files of the Secretary of Agriculture; and memorandum, Julius Klein to Hoover, July 22, 1921, Herbert C. Hoover Papers, Herbert Hoover Presidential Library, West Branch, Iowa, 1-I/2.

[20] Leon M. Estabrook to Wallace, March 14, 1921; George Livingston to Wallace, April 22, 1921, BAE Files; W. F. Callander and G. F. Warren to Taylor, October 20, 1921, Files of the Secretary of Agriculture; and Taylor to Louis G. Michael, September 19, 1922, Taylor Papers.

[21] Leon M. Estabrook to Julius Klein, October 31, 1921; Julius Klein to Hoover, October 17, 1921, Files of the Bureau of Foreign and Domestic Commerce, File 155, Box 926; and Julius Klein to Leon M. Estabrook, November 5, 1921, *ibid* and BAE Files.

of a joint committee on statistics, furthermore, represented a permanently successful effort at working together on the collection of data in the domestic field.[22]

But jurisdictional boundaries were nebulous and interdepartmental cooperation at best precarious. While Klein was praising the Agriculture Department's instructions to its foreign commissioners, he was lamenting the "absurdity of . . . duplication of our work" and the unwarranted meddling in "our business" by the same department. Hoover was likewise more than ever convinced that commercial attachés were "in a position to collect . . . [agricultural] data with perhaps only casual guidance from agricultural experts in Washington." [23] On the other hand, in April 1921 Wallace had told a grain marketing association that if farmers were effectively to adjust their production to existing demand, they had to have complete data on conditions in competing countries. "The Department of Agriculture," he assured his audience, "should furnish this information, both as to the supply and the demand." Declaring that marketing work in farm commodities "must be carried on by those who have a sympathetic attitude toward the producer as well as the consumer," Taylor insisted that it would be ill-advised for a "non-agricultural department or bureau" to undertake such service. "The De-

[22] Edward F. Feely to Julius Klein, December 5, 1921; D. S. Bullock to Edward F. Feely, December 3, 1921; Leon M. Estabrook to Edward A. Foley, October 31, 1921; C. F. Herring to Julius Klein, October 30, 1921; Wallace to Hoover, December 9, 1921, Files of the Bureau of Foreign and Domestic Commerce, File 155, Box 926, File 261.1, Box 1453, and File 371, Box 1867; Julius Klein to Leon M. Estabrook, December 15, 1921; Joint Committee on Statistics to Wallace and Hoover, March 28, 1922; Wallace to Hoover, April 20, 1922, BAE Files; Wallace to Hoover, October 17, 1921, Files of the Secretary of Agriculture; and W. A. Jump to [?] Taylor, May 17, 1921, Hoover Papers, 1-I/2.

[23] Julius Klein to Edward F. Feely, November 30, 1921; Hoover to Charles G. Dawes, November 29, 1921, Files of the Secretary of Commerce, File 80553, Box 485; and Julius Klein's penciled note at the top of Wallace to Hoover, July 14, 1921, Hoover Papers, 1-I/2.

partment of Agriculture," he flatly asserted, "is regarded as the source of information on all subjects relating to agriculture." [24] And so the lines were drawn.

Control over foreign agricultural reporting became a central issue in the discussions on executive reorganization. Before Wallace took office, the question of duplication of work between the Departments of Agriculture and Commerce had been under consideration in Congress and was soon to become a prime concern of the recently created Joint Committee on Reorganization. Interest focused primarily on Agriculture's Bureau of Markets, later incorporated into the BAE, and a lively controversy arose over placement of this agency.[25]

Wallace, Hoover, and Walter F. Brown, chairman of the joint committee, met in July 1921 to discuss reorganizational matters relating to the Agriculture and Commerce Departments. Though the conference settled nothing, Brown's suggestion that the two secretaries submit statements of their views to his committee was accepted. After the meeting, Wallace reported to Harding that he was irritated to learn that Brown had already formed "some rather definite opinions," one of which was that the Bureau of Markets should be transferred to the Commerce Department. In his statement to the joint committee a short time later, the secretary asserted that such a move would be "indefensible from an administrative standpoint." Applying his progressive views on expertise, Wallace explained that "the science of the horticulturist, plant physiologist, plant pathologist, refrigeration expert, and statistician, all must be brought to bear upon the marketing problem." Since the Agriculture Department already provided ser-

[24] Address before the Farmers' Grain Marketing Committee, Chicago, April 6, 1921, copy in Taylor Papers; and Taylor to Wallace, July 16, 1921, BAE Files.

[25] Memorandum, H. F. Fitts to Bureau Chiefs, March 10, 1921, BAE Files; and Henry C. Taylor, "A Farm Economist in Washington," p. 73, ms. in Taylor Papers.

vices along these lines, it was only logical that the Bureau of Markets should remain where it was. Its transfer would merely necessitate the creation of similar functions in the Commerce Department and thus result in the duplication of work which reorganization was supposed to eliminate.[26]

Beyond the "technical" argument, however, was a more compelling reason for retention of the bureau. In Wallace's eyes Hoover and the Commerce Department represented business-financial interests, the very elements which were antagonistic toward agriculture. Given responsibility for the collection and dissemination of agricultural information, Wallace believed, Hoover's department would surely sacrifice the interests of farmers to those of business. According to a manuscript drawn up by the BAE, Commerce had "no intimate contact with the American farmer"; worse yet, its representatives possessed a "completely economic and commercial point of view." [27] Under this agency, Taylor maintained, services in agricultural marketing "would be submerged by the overwhelming industrial and commercial interests in the Department of Commerce." The work of the Bureau of Markets, therefore, had to be carried on not only by experts in the field of agriculture but by men with an agrarian viewpoint and a sympathetic understanding of the problems of farmers.[28]

Hoover delayed until October 1921 before forwarding his statement to Brown. Meanwhile, he had a member of his staff study the activities of the Bureau of Markets to determine if the contemplated transfer would be advisable. The report assured the secretary that "practically all these functions could be performed

[26] Wallace to Harding, August 8, 1921; Wallace to Walter F. Brown, January 16, 1923, Files of the Secretary of Agriculture; see also Charles G. Dawes, *The First Year of the Budget of the United States* (New York: Harper and Brothers, 1923), p. 23.

[27] "Who Should Handle Foreign Agricultural Work?," ms. in Taylor Papers.

[28] Taylor, "A Farm Economist in Washington," p. 75.

by the Department of Commerce in connection with its other activities of similar nature. . . ." [29] This advice both aided Hoover in making up his mind and served to confirm a view he had held before becoming Secretary of Commerce. Obviously referring to the Bureau of Markets in his telegram accepting appointment to the post, he had indicated to Harding that "there must be included in it [the Commerce Department] such bureaus as properly belong to its field and I trust that I should have a voice and your support in the reorganization." [30]

In his detailed memorandum to the chairman of the joint committee, Hoover unequivocally expressed his views. The Commerce Department, he maintained, had been created "to foster commerce in its most comprehensive sense . . . ," which meant that "the functions of the Department of Agriculture should end when production on the farm is complete and the movement therefrom starts, and at that point the activities of the Department of Commerce should begin." Answering Wallace's argument that his staff was better trained to handle foreign agricultural work, Hoover countered that scientific research and technical examination entered into the marketing and distribution of farm commodities only incidentally, particularly as compared to commercial research and investigation. As far as Hoover was concerned, the functional division between the two departments was too obvious to question: "The Department of Agriculture should tell the farmer what he can best produce, based on soil, climate and other cultural conditions, and the Department of Commerce should tell him how best to dispose of it." [31]

[29] E. G. Montgomery to Hoover, September 7, 1921, Files of the Bureau of Foreign and Domestic Commerce, File 155, Box 926.

[30] Telegram, Hoover to Harding, February 24, 1921, Harding Papers, Box 698.

[31] Hoover to Walter F. Brown, October 20, 1921; and memorandum, Hoover to Walter F. Brown, October 20, 1921, Files of the Secretary of Commerce, File 80553, Box 485 and Taylor Papers.

Later Hoover was to claim that he had not endorsed the transfer of the Bureau of Markets, that in fact he had opposed it. Maintaining that the memorandum of October 1921 had merely outlined a "theoretical case," he denied that his intention had been to convince Brown to recommend Commerce jurisdiction over the bureau. "While a good paper case could be made in logical arrangement of government functions and economies for [the transfer] . . . ," the secretary recalled, "I felt that the Bureau had had large relations direct to the farmer, and I strongly advised against the transfer on public grounds." [32] Clearly sensitive over the issue, Hoover and his staff in answer to inquiries on the subject denied any desire to transfer the division, going so far as to inform one correspondent that "there has never been any discussion of moving the Bureau of Markets to the Department of Commerce." [33]

Hoover's amazingly convenient memory, however, had allowed him to forget the strong message of his October 1921 statement. Though he did not specifically recommend transfer of the Bureau of Markets, the implication of the theoretical case developed in the memorandum could not have been lost on anyone familiar with the circumstances of reorganization. In December 1921, moreover, Hoover wrote to Brown again, suggesting "for the tail end of your formula which we discussed yesterday, the following wording: 'In the commercial stage shall be limited to the inspection, grade, and investigation of transportation and warehousing equipment of perishable products.' " While he did

[32] Hoover to Howard M. Gore, November 24, 1924; Walter F. Brown to Hoover, July 19, 1924, Coolidge Papers, File 1, Box 1; see also Louis G. Michael to Taylor, May 28, 1924, Taylor Papers. Taylor maintained that if Hoover had no desire to transfer the Bureau of Markets, "his subordinates were thinking pretty definitely along those lines. . . ." Taylor to Theodore Saloutos, January 24, 1940, *ibid.*

[33] Hoover to R. N. Wilson, January 19, 1922; Hoover to Ralph P. Merrill, January 11, 1922; and Richard S. Emmet to editor of *Washington Star,* August 28, 1923, Hoover Papers, 1-I/2 and 1-I/258.

not name the department to which the "formula" was to be applied, there can be no doubt that Agriculture was the one in question. The Commerce secretary also offered to work out "formulas" for his department and "for the Bureaus that have been considered as appropriately coming to this Department." [34] Despite later denials, therefore, Hoover had more than a theoretical interest in the relationship of the two departments and the location of the Bureau of Markets.

Like the controversy over the Forest Service, the possible transfer of the Bureau of Markets resulted in a flood of letters into the Agriculture Department. Most vocal in opposition to the move, all of the major farm organizations wrote to Wallace strongly supporting his stand for retention of the bureau. Agricultural colleges, forestry associations, farm editors, as well as individual farmers, likewise rallied behind the secretary. Highly pleased with the response, Wallace urged all of his supporters to make their views known to Brown and the Joint Committee on Reorganization. The Joint Commission of Agricultural Inquiry also took up the matter, but, notwithstanding Chairman Sidney Anderson's sympathy toward Wallace's views, the committee was unable to agree and dropped the question without making a recommendation. [35]

As the controversy dragged on, Wallace grew increasingly irritated. A request from Brown in December 1922 for comments on Hoover's statement of October 1921 tended to sharpen his feelings. It was the first that Wallace had seen of the Com-

---

[34] Hoover to Walter F. Brown, December 3, 1921, *ibid.*, 1-I/257.

[35] See Files of the Secretary of Agriculture; and Henry C. Taylor, "Henry C. Wallace and the Farmers' Fight for Equality," p. 4, ms. in Taylor Papers. Later there was disagreement within the Farm Bureau over which department to support. The problem was solved by a resolution recommending that both departments work to expand exports and report agricultural information. L. G. Michael to Taylor, undated; O. E. Bradfute to Chester C. Davis, February 21, 1924; and L. G. Michael to Taylor, June 24, 1924, *ibid.*

merce chief's memorandum, and he demanded to know why it had not been called to his attention earlier. Noting that Hoover had been "exceedingly frank in stating his views and [that] there is no possibility of misunderstanding them," he challenged his colleague's contention that the Agriculture Department was not intended to operate in the field of marketing and distribution by calling upon historical precedent. In arguing that commercial aspects of farming had always been considered within Agriculture's jurisdiction, he pointed out that the department had engaged in such activities since its creation in the middle of the nineteenth century. To transfer these functions to another agency, the secretary reiterated, would be "indefensible from every standpoint of sound administration. . . ." If Hoover's advice were followed, the Commerce Department would have to bring together a team of farm experts "which would rival in number the group now existing in the Department of Agriculture and . . . result in vast duplication of work. . . ." As for any functional overlapping which then existed between the two divisions, Wallace asserted that it had "grown out of the persistent encroachment by the Department of Commerce upon the fields properly belonging to the Department of Agriculture." [36]

Neither Wallace's impassioned appeals nor the protests of farmers and their spokesmen could sway Brown, who remained convinced of the logic of Hoover's theoretical case. Had Harding not intervened, the chairman most certainly would have recommended transfer of the Bureau of Markets. The President admitted, with an admirable bit of candor, that he had "never been able to judge the merits of the controversy," but conceded that he could see "some logic in this Bureau being located in the Department of Commerce. . . ." As in the case of the proposed transfer of the Forest Service, though, his first concern was with

[36] Wallace to Walter F. Brown, January 16, 1923, Files of the Secretary of Agriculture.

maintaining harmony in the cabinet. Because of Harding's feeling that this could best be accomplished by leaving the departments largely as they were, Brown abandoned his original plan and no mention of the Bureau of Markets appeared in his tentative reports of January 1922.[37] The cabinet recommendations to the joint committee in February 1923 likewise avoided the issue, and the committee, following closely administration suggestions, left untouched the sensitive questions of executive reorganization in its proposed bill of the following year.[38]

When it became clear that he would not secure his purpose through reorganization, Hoover stepped up expansion of the Commerce Department's foreign work in agricultural reporting. At Harding's request, Congress granted the department a special appropriation of $500,000 in March 1923 to investigate foreign sources of raw materials not available or in short supply in the United States. The President's recommendation had proposed only a study of crude rubber, but friends of agriculture in Congress demanded that fertilizer nitrates and binder-twine sisal also be included. Debate on the question resulted in a rather ambiguous wording of the appropriation act, authorizing the Commerce Department "to investigate related problems in the development of the foreign trade of the United States in agricultural and manufactured products." [39] Little did agrarian representatives real-

[37] Harding to Halvor Steenerson, May 27, 1922, Harding Papers, Box 2; *Washington Herald,* March 11, 1922; Harding to Walter F. Brown, February 13, 1923, reprinted in United States Congress, Senate, *Reorganization of the Executive Departments,* Document 302, 67th Cong., 4th Sess., pt. iii; and ms. of Brown's statement to the joint committee, January 21, 1924, Hoover Papers, 1-I/257. Brown remembered later that he had dropped the transfer proposal upon Hoover's advice. This, as noted, seems highly improbable. Walter F. Brown to Hoover, July 19, 1924, Coolidge Papers, File 1, Box 1.

[38] *Reorganization of the Executive Departments,* p. 3.

[39] *Congressional Record,* 67th Cong., 4th Sess., pp. 3734, 4686–88; G. F. Leonard to Taylor, March 21, 1924, Taylor Papers; and *Annual Report of the Secretary of Commerce,* 1923, p. 114.

ize that in securing this change they were opening the way for further encroachment upon the domain claimed by Agriculture.

Actually, the crude rubber investigation, the principal project for which the special appropriation had been intended, represented a noteworthy example of cooperation between the Agriculture and Commerce Departments. This happy situation resulted from a general agreement on assignment of functions. The Department of Commerce undertook to discover the nature and extent of international combinations in control of the world's crude rubber supplies and, if possible, to find alternative sources.[40] Concentrating on the purely technical aspects of the project, Wallace's department conducted a systematic exploration of possible rubber producing areas in Central and South America with a view toward encouraging the development of new sources. In addition, the Bureau of Plant Industry experimented with various species of rubber trees, searching for one which might be adapted to conditions in the United States.[41] There was no conflict over the division of labor on the survey largely because Wallace readily accepted the scientific problems involved in rubber production as the limitation of his department's activities. Such an amicable arrangement was possible only because the investigation concerned a commodity not produced by American farmers.[42] Had the subject been wheat,

[40] *Annual Report of the Secretary of Commerce,* 1923, p. 35; and Julius Klein to Willam H. King, December 19, 1925, Files of the Bureau of Foreign and Domestic Commerce, File 621.2, Box 2947.

[41] Wallace to Lee S. Overman, May 20, 1924, Files of the Secretary of Agriculture; and P. L. Palmerton to E. Burris Warner, March 2, 1923, Files of the Bureau of Foreign and Domestic Commerce, File 621.2, Box 2947.

[42] Hoover to Wallace, April 21, 1923; Wallace to Hoover, May 28, 1923, Files of the Secretary of Agriculture; and C. W. Pugsley to Hoover, March 30, 1923, Files of the Bureau of Foreign and Domestic Commerce, File 370, Box 1865.

corn, or dairy products, Wallace would not have relinquished so quickly the Agriculture Department's right to participate in the commercial aspects of the study.

This spirit of cooperation, however, did not extend into the larger project initiated by the Commerce Department under authority of the special appropriation act. Shortly after passage of the act, Hoover began laying plans for a massive investigation into agricultural export markets. The first step was the creation of an advisory commission to give "definition and direction" to the project, which, as the Commerce secretary took care to make clear, would "actually be carried out by men in the Department." So that the undertaking would appear to be an interdepartmental effort, Hoover contacted Wallace in regard to a representative from Agriculture. Wallace suggested Taylor, and when the commission was announced the BAE chief was among those assigned to the group, which also included representatives of Congress, major farm organizations, financial interests, commodity associations, and food processors, as well as the Commerce Department itself.[43]

Held on March 24, 1923, the first and, as it resulted, the only meeting of the Commission on Agricultural Exports went pretty much as Taylor had expected. "He [Hoover] talked as if the whole field of production was the field for the Department of Agriculture," the bureau chief reported to Wallace, "and that marketing was a field for the Department of Commerce." Project head Frank M. Surface of the Commerce Department presented a previously formulated plan, which the conference promptly approved, and after a short discussion the meeting adjourned. It

[43] Hoover to Wallace, March 7, 1923; Wallace to Hoover, March 9, 1923, Files of the Secretary of Agriculture and Taylor Papers; telegram, Hoover to George H. McFadden, March 10, 1923; and telegram, Hoover to C. W. Hunt, March 10, 1923, Hoover Papers, 1-I/5.

was evident from the start that Surface would direct the work and that the commission was to have little function other than to lend its influence to the investigation.[44]

The Survey of World Trade in Agricultural Products, as the investigation was called, proceeded along two lines. One part was a statistical study of imports and exports of the United States and of other important agricultural producing countries, the main purpose of which was to determine the trend of international trade in the postwar period and to obtain accurate information on farm production and consumption throughout the world. Secondly, the investigation analyzed production costs and marketing practices in various countries, with special emphasis on systems of credit, finance, storage, warehousing, and distribution. Included in this section was a study of tariff policies and general economic conditions as they affected world trade in farm goods. Beyond the specific objectives, Hoover anticipated that the Commerce Department's survey would lay the basis for the formulation of a sound and enduring agricultural policy by the federal government.[45]

All of the data-gathering and the major part of the analysis for the trade investigation was carried on by officials of the Department of Commerce. Surface kept the advisory commission informed on the progress of the work, but in no instance did he solicit advice on procedure or content of the study. After the initial conference, contact with the group was on the basis of cor-

[44] F. M. Surface to Julius Klein, March 26, 1923, Files of the Bureau of Foreign and Domestic Commerce, File 370, Box 1865; Taylor to Wallace, April 18, 1923; Taylor's report of the conference, March 24, 1923, Taylor Papers; and "Investigation of World Trade in Agricultural Products," ms. in Hoover Papers, 1-I/5.

[45] Julius Klein to Malcolm Stewart, April 7, 1923; Frank M. Surface to Chester Lloyd Jones, May 22, 1923, Files of the Bureau of Foreign and Domestic Commerce, File 370, Box 1865; memorandum, F. M. Surface to Julius Klein, March 14, 1923, Hoover Papers, 1-I/5; and *Annual Report of the Secretary of Commerce,* 1923, pp. 114–16.

respondence between the Commerce Department and individual members.[46] Outside of Taylor's token presence on the commission, the Agriculture Department took no part in the survey. The investigation was, in short, a Commerce project, and the commission served only to make it appear a joint effort.

Beginning in January 1924, the Commerce Department published its findings in a series of *Trade Information Bulletins* under the title of "Survey of World Trade in Agricultural Products." The first bulletin analyzed the distribution of American farm exports, and subsequent numbers concentrated on specific commodities and problems involved in the export trade. Totaling 14 in all, the publications dealt with such subjects as merchandizing of wheat, transportation in relation to foreign trade, import duties on farm goods, and foreign marketing of various agricultural products.[47] Material contained in these studies, however, was in many cases a duplication of information either made available previously or in preparation by the BAE. In connection with its expanded farm economics program, the Agriculture Department devoted the *Yearbook of Agriculture* in the years 1921 through 1925 to studies on the business aspects of farming, several of which closely paralleled parts of Hoover's investigation. While providing a wealth of information on trade in farm products, furthermore, the Commerce Department's survey was of no greater practical use to individual producers than that distributed by the BAE. In both cases farmers were simply ill-equipped to translate these studies into programs of production adjustment and improved methods of marketing.

The Survey of World Trade in Agricultural Products repre-

[46] See Files of the Bureau of Foreign and Domestic Commerce, File 370, Box 1865 for correspondence relating to the investigation.

[47] H. M. Strong, *Distribution of Agricultural Exports from the United States* (Department of Commerce Trade Information Bulletin 177) and subsequent bulletins in the series; and *Annual Report of the Secretary of Commerce*, 1924, p. 118.

sented a marked expansion of the Commerce Department's activities in the area of farm marketing. Until the spring of 1923 encroachment on the field of foreign work claimed by the BAE had been irritating but not considered serious by Agriculture officials.[48] But Hoover's vigorous program under authority of the special appropriation act, with its duplication of functions currently undertaken by the bureau, greatly alarmed Wallace and members of his department. As a result, relations between Agriculture and Commerce, while never cordial, deteriorated and set the background for a bitter interdepartmental fight in early 1924.

In late 1923 Louis G. Michael, a consulting specialist with the BAE, wrote to Taylor that "a desperate attempt is going to be made to do something this next session of Congress." He was not in possession of the details, but there was to be some kind of move by the Commerce Department to take over completely all foreign agricultural marketing work.[49] Early the following year the nature of Hoover's plan was revealed. Samuel E. Winslow, a Massachusetts representative, sponsored a bill to establish, among other things, a foreign service in the Department of Commerce which would be the sole agency for the investigation and reporting of economic and commercial aspects of agriculture abroad.[50] As might well be expected, the Commerce Department had encouraged the introduction of the measure and had supplied Winslow with a good deal of technical information to aid in its drafting. Since most of the department's foreign service work had been based on authority granted in its annual and special appropriation acts, Hoover was anxious to have these activities placed on a permanent statutory basis. Not only would the Winslow bill provide for this, but it would also serve the larger

---

[48] Taylor, "A Farm Economist in Washington," p. 76.
[49] Michael to Taylor, November 11, 1923, Taylor Papers.
[50] *Congressional Record,* 68th Cong., 1st Sess., p. 569.

purpose of preventing Agriculture from performing functions which Hoover believed properly belonged to his office.[51]

The reaction of Wallace's department was to counter with a proposal of its own. Introduced into the House in January 1924 by John C. Ketcham of Michigan, the measure provided for expansion of the Agriculture Department's foreign work and would have given diplomatic status to its representatives abroad.[52] As with Commerce, Agriculture derived most of its authority for activities outside of the United States from annual appropriation acts. When Wallace first took office, an advisor called his attention to the rather precarious basis for foreign work and recommended a revision of the fundamental law to provide permanently for activities in this area.[53] The Ketcham bill was designed both to make the necessary revisions and to meet the challenge of the Commerce Department's Winslow bill.

Immediately after the Winslow bill was introduced, the Commerce Department set to work to rally support for the measure. In collaboration with Winslow, the department enlisted a group of sympathetic witnesses to testify in behalf of the legislation before the House Committee on Interstate and Foreign Commerce, making every effort to secure prominent men from influential organizations.[54] Instructing its district offices throughout the

[51] O. P. Hopkins to B. C. Getsinger, January 4, 1924; and Julius Klein to Samuel E. Winslow, Janury 19, 1924, Files of the Bureau of Foreign and Domestic Commerce, File 127, Box 815. The Winslow bill also provided for the transfer of the economic and commercial work of the Consular Service to the Commerce Department. This aspect of the measure met with opposition from the State Department. Julius Klein to Charles Lyon, January 8, 1924, *ibid.*

[52] *Congressional Record*, 68th Cong., 1st Sess., p. 1071.

[53] E. D. Ball to George Livingston, March 14, 1921; and C. L. Luedtke to Leon M. Estabrook, January 11, 1922, BAE Files.

[54] O. P. Hopkins to B. C. Getsinger, January 4, 1924; Samuel E. Winslow to Julius Klein, January 23, 1924; and Julius Klein to Samuel E. Winslow, January 24, 1924, Files of the Bureau of Foreign and Domestic Commerce, File 127, Box 815.

country to contact business and financial interests, local chambers of commerce and divisions of the National Manufacturing Association, editors, and others, the Bureau of Foreign and Domestic Commerce launched a program to generate a petition and letter-writing campaign in support of the Winslow bill. The Commerce Department also supplied the House committee with information strengthening its claim to jurisdiction over the economic and commercial aspects of foreign work in farm products.[55] The fervor and urgency with which the Commerce Department sought to secure passage of the Winslow bill illustrated the importance it placed on obtaining complete control of the agricultural export field.

The interest of the Agriculture Department in the Ketcham bill was no less intense. "We regard it," Wallace wrote to a member of the House, "as necessary to enable us to continue unhampered our service to American agriculture." The secretary explained that the legislation defined the scope of his department's foreign functions so clearly that "there can not possibly be any misunderstandings as to the authority for work which the Department has been carrying on for the past forty years." Wallace assigned Louis G. Michael of the BAE to the task of contacting congressmen and farm organizations with the purpose of securing support for the bill. Chester C. Davis, a Montana state agricultural commissioner who had been engaged by the department to lobby for another measure, also worked to rally backing for the Ketcham bill.[56] Both Agriculture and Commerce, therefore,

[55] See, for instance, O. P. Hopkins to B. C. Getsinger, January 4, 1924; W. R. Rastall to New York District Office, January 26, 1924; telegram, Hopkins to B. C. Getsinger, January 30, 1924; Julius Klein to Samuel E. Winslow, March 6, 1924; and other correspondence in *ibid.*

[56] Wallace to Gilbert N. Haugen, February 23, 1924, Files of the Secretary of Agriculture; Michael to Taylor, March 28, 1924; Michael to Taylor, February 28, 1924; Michael to Taylor, May 28, 1924; and Chester C. Davis to Taylor, November 19, 1924, Taylor Papers. Davis was working to secure passage of the McNary-Haugen bill, a subject taken up in the next chapter.

were determined to resolve their conflict of three years by eliminating the ambiguity that permitted encroachment on areas
which they were confident fell within their respective jurisdictional spheres.

Besides laboring to obtain enactment of the Ketcham bill,
Wallace's department also sought to eliminate objectionable
parts of the Winslow bill. After studying the measure, the solici-
tor of the Agriculture Department warned Wallace that the section of the bill placing authority in the Commerce Department
for the investigation and reporting of economic and commercial
aspects of agriculture would virtually remove the BAE from this
field of activity. The secretary, of course, was fully aware of this
from the outset and had already petitioned Winslow and the
Committee on Interstate and Foreign Commerce to have the section stricken before the measure was reported back to the
House.[57] Naturally set against any such revision, Hoover used
his influence with the committee to block any changes in the
legislation.[58] In May 1924 Wesley L. Jones of Washington introduced a companion bill to Winslow's in the Senate, but with
several modifications which were purportedly to meet the objections of the Agriculture Department. Wallace made clear to
Jones, however, that his measure was no more satisfactory than
the House version and urged changes "to remove all doubt as to
the relative functions of the Departments of Agriculture and
Commerce in the foreign field." [59]

[57] R. W. Williams to Wallace, March 20, 1924, *ibid;* and Wallace to
Samuel E. Winslow, February 29, 1924, Files of the Secretary of Agriculture.

[58] Michael to Taylor, March 3, 1924; Michael to Taylor, March 28,
1924, Taylor Papers; and United States Congress, House, Committee on
Interstate and Foreign Commerce, *Hearings on the Winslow Bill,* 68th
Cong., 1st Sess., pp. 104–8.

[59] *Congressional Record,* 68th Cong., 1st Sess., p. 9393; Julius Klein
to Charles L. Chandler, May 27, 1924, Files of the Bureau of Foreign
and Domestic Commerce, File 127, Box 815; Wallace to W. L. Jones,
June 2, 1924, Files of the Secretary of Agriculture; and Michael to
Taylor, May 28, 1924, Taylor Papers.

As the two departments stepped up their campaigns, various interest groups joined in the fight. Alignment on the measures followed established patterns, as groups traditionally connected with the Commerce Department backed the Winslow bill and those with ties to Agriculture favored the Ketcham bill. Business, manufacturing, and financial interests along with their spokesmen responded to Hoover's call and endorsed the Commerce measure, while farm organizations, marketing associations, and agricultural colleges generally favored the Ketcham bill.[60] Although President Coolidge remained outside of the interdepartmental fight, it was understood that he agreed with Hoover and wished to see the Winslow measure enacted.[61]

Within a short time the Ketcham bill emerged from the sympathetic House Committee on Agriculture and Forestry with a favorable report. In April the measure was brought to the floor and quickly passed with a minimum of discussion.[62] The legislation's reception in the Senate, however, was less enthusiastic. "When it comes to a question between the interests of Commerce and Agriculture," Michael reported to Taylor, "you can easily see that we are practically blocked [in the Senate] without outside pressure being brought to bear." Michael redoubled his efforts to line up support for the bill and secured commitments from several senators. But, as he complained to Taylor, the problem was in getting it reported out of the Committee on Agriculture and Forestry.[63] In a letter to committee chairman George W. Norris,

[60] See Files of the Bureau of Foreign and Domestic Commerce, File 127, Box 815, Files of the Secretary of Agriculture, and Taylor Papers. Although the Farm Bureau officially supported the Ketcham bill, there was a faction within its membership which favored the Winslow bill because of the belief that the Commerce Department was better equipped to dispose of farm surplus abroad. Michael to Taylor, June 7, 1924, Taylor Papers.

[61] Winslow claimed his bill had Coolidge's support. *Congressional Record*, 68th Cong., 1st Sess., p. 10725.

[62] *Congressional Record*, 68th Cong., 1st Sess., p. 6500.

[63] Michael to Taylor, May 15, 1924; and Michael to Taylor, May 28, 1924, Taylor Papers.

Wallace requested that the Ketcham bill be rushed through with no hearings. "We would appreciate it if it could be handled in this way," the secretary urged, "as we are very anxious to get it reported out and on the way to passage." [64] At the time the committee was holding hearings on legislation for the creation of a public power project at Muscle Shoals, a matter which would dominate Norris' attention until well into the 1930's, and the Nebraska senator refused to open the way for anything that might interfere. As a result, the Ketcham bill was held up in committee until the end of the session.

The Winslow bill fared even less well. After extended hearings, it was reported out of the Committee on Interstate and Foreign Commerce and placed on the House calendar. The first session of the Sixty-eighth Congress was drawing rapidly to a close, however, and it was clear that time would run out before the bill came up. Supporters of the measure attempted to secure a suspension of the rules to bring it to a vote out of order, but failed to obtain the necessary two-thirds majority.[65] In the meantime, the Jones bill died in the Senate committee.

The fight over the Winslow and Ketcham bills left the split between Commerce and Agriculture wider than ever. Antagonism between Wallace and Hoover had grown to the point where any possibility of their working together for the good of agriculture, if indeed such a possibility ever existed, had completely disappeared. After the close of Congress, Michael prepared a detailed statement tracing the alleged encroachment of the Commerce Department upon the area of farm marketing, and Wallace released the document in a final effort to gather support for Agriculture's case.[66] Furious at this appeal to the public, Hoover

[64] Wallace to George W. Norris, April 18, 1924, Files of the Secretary of Agriculture.

[65] *Congressional Record*, 68th Cong., 1st Sess., p. 10729.

[66] "Encroachment of the Department of Commerce upon the Department of Agriculture in Marketing and Agricultural Economics Investigations, July, 1924," ms. in BAE Files and Taylor Papers; and Michael to

charged that "publication of the memo . . . was a gross violation of the Cabinet confidence and deliberately intended to mislead." [67] The conflict also strained relations between the Agriculture Department and the President's office. Later in the year Coolidge's secretary, without comment, sent Wallace a copy of a memorandum which the President had received complaining of Michael's efforts to insure the defeat of the Winslow bill if it were brought up in the next Congress. "It might be proper to inquire," noted the message, "as to how far a Civil Service appointee paid out of Federal treasury funds may go in lobbying against a measure sponsored by a member of the President's cabinet, and to what extent he may identify himself with men who are the self declared opponents of the President and his policies." [68]

Soon after Wallace's death in October 1924, the Commerce Department resumed its efforts to eliminate Agriculture from foreign marketing work. In November Hoover appealed to acting Secretary of Agriculture Howard M. Gore to remove the department's agents abroad who were "engaged in work parallel with and in duplication of that carried on systematically by the Department of Commerce." A month later he wrote to Coolidge recommending the withdrawal of foreign agricultural representatives who were carrying on marketing work. Although the President took no direct action to resolve the jurisdictional dispute, he offered Hoover the post of Secretary of Agriculture. But the Commerce chief declined the appointment, as he later remem-

George N. Peek, January 7, 1925, George N. Peek Papers, Western Historical Manuscripts Collection, University of Missouri Library, Columbia.

[67] Hoover to George N. Peek, December 19, 1924; Hoover to Milton W. Shreve, May 22, 1924 (not sent), Hoover Papers, 1-I/258; and Julius Klein to Julian Arnold, July 1, 1924, Files of the Bureau of Foreign and Domestic Commerce, File 127, Box 815.

[68] C. B. Slemp to Wallace, October 20, 1924 and attached memorandum, Charles E. Hearst Papers, University of Northern Iowa Library, Cedar Falls.

bered, "on the ground that I could do more for the farmers as Secretary of Commerce and was not a technologist on agricultural production." [69]

In January 1925 one newspaper speculated that the new Secretary of Agriculture "would be found in a man in accord with Secretary Hoover, not because the President is permitting the commerce head to do his cabinet picking for him, but because he and Mr. Hoover think alike on the farm question." [70] If this assessment were correct, the man chosen fit the qualifications perfectly. Formerly the president of Kansas State College, the new Agriculture head, William M. Jardine, agreed with Hoover and Coolidge in every particular. So satisfactory was he, in fact, that Julius Klein could write shortly after his appointment: "The present Secretary of Agriculture is very friendly to us, and I am confident that we can work out a scheme for effective collaboration between their foreign service and ours so as to eliminate practically all possibility of overlapping of function." Michael later confirmed this appraisal when he complained to Taylor that Jardine was most sympathetic to the views and wishes of the Commerce Department. [71]

The activities of the Agriculture Department after Jardine took over clearly testified to the fact that the new secretary thought differently from Wallace. Not only was the expansion of foreign marketing halted, but agents of the department gradually ceased engaging in functions claimed by Commerce. Undoubtedly the most pointed evidence that Agriculture had come under new leadership was the dismissal of Taylor. In January 1925 a

[69] Hoover to Howard M. Gore, November 24, 1924; Hoover to Coolidge, December 20, 1924, Coolidge Papers, File 1, Box 1; and Hoover, *Memoirs*, vol. 2, p. 111.

[70] *Philadelphia Public Ledger*, January 19, 1925, p. 4.

[71] Julius Klein to Richard A. May, April 15, 1925, Files of the Bureau of Foreign and Domestic Commerce, File 127, Box 815; Michael to Taylor, March 17, 1926; and Michael to Taylor, October 14, 1927, Taylor Papers.

Philadelphia newspaper had predicted that there would be an "attempt to break down the bureaucratic clique which is alleged to have employed the official agencies of the [Agriculture] department to combat Secretary Hoover's policies on farm relief." [72] As Wallace's principal advisor on agricultural economics, Taylor was the key member of that clique, and, to the surprise of no one, six months after the new secretary assumed office he abruptly fired the BAE chief without explanation.[73] The reasons for Jardine's action, however, were clear. At the head of a devoted staff carefully trained over the past four years, Taylor represented the main obstacle to the Commerce Department's domination of the field of foreign agricultural marketing, and thus he had to go. Embittered by his dismissal, the bureau chief assured Henry Agard Wallace that "regardless of the press notices I did not resign." The farm editor saw the move as merely one more step in the "Hooverizing of the Department of Agriculture," which in effect "told farm folks who have been striving for equality for agriculture to 'go home and slop the pigs.' " [74]

The fight between Agriculture and Commerce stemmed from three causes. First of all, Wallace and Hoover were predisposed to conflict because of their difficulties before entering the Harding administration. The bitter disagreement over policies of the Food Administration during the war and Wallace's scathing campaign against Hoover's presidential candidacy in 1920 left scars which were bound to affect their relationship in the cabinet. Neither man, furthermore, possessed a temperament which would have allowed him to rise above these earlier differences. Thus

[72] *Philadelphia Public Ledger,* January 19, 1925, p. 4. Taylor also expected a purge of the department when the new secretary took office. Taylor to H. A. Wallace, February 6, 1925, Taylor Papers.

[73] William M. Jardine to Taylor, August 10, 1925, Taylor Papers.

[74] Taylor to H. A. Wallace, September 7, 1925, *ibid.;* and *Wallaces' Farmer* 50 (August 28, 1925): 1090.

when the issue over foreign marketing arose, both ruled out almost automatically the possibility of compromise and cooperation—they were antagonists and would conduct their interdepartmental war on that basis.

The controversy, moreover, was of a type common to bureaucratic administration, particularly where jurisdictional lines are ill-defined. Neither side was in fact encroaching upon the activities of the other, because responsibility for foreign agricultural marketing had never been clearly established. Both departments, to be sure, had engaged in this field of work for sometime, but under the authority of temporary appropriation acts and not of their respective organic laws. Until 1921 activity along these lines was limited and there had arisen no occasion for disagreement. But, with the expansion of Agriculture and Commerce into the same area after the war, conflict naturally resulted. Based upon their previous activities, both agencies laid claim to jurisdiction over the contested sphere and the inevitable bureaucratic power struggle ensued.

Another, and perhaps more fundamental, cause of the controversy arose from a characteristic ingrained in American development. As the United States moved from an agrarian to an industrial economy, antagonism between agricultural and business interests grew. The struggle between Wallace and Hoover was, in part, a manifestation of this process. Secretary Wallace narrowly identified with the farmers, and he viewed with suspicion industrial-financial interests and their spokesmen, of which he considered the Commerce Department the prime example. Though certainly more broadly oriented in his thinking, Hoover nevertheless held an unmistakable disdain for the agrarian and a decided leaning in favor of business groups. At least one outside observer recognized the more basic problem involved when he noted that "this thing has gone far beyond any personal controversy between the two Secretaries. They are simply agents expressing the conflicting aims of big business and organized farmers." Similarly, Michael insisted that the struggle was not really interdepartmental, but part of "the great conflict between organized industry on the one side and agriculture['s] attempt to organize on the other. . . ." [75] While this aspect of the dispute should not be overemphasized, Wallace and Hoover reflected the attitudes of contrasting elements in society, a factor which served to sharpen the lines of their contest over foreign marketing work.

Whatever the reasons for the fight, the country's farmers, the very group both secretaries were endeavoring to serve, were the ones to suffer. Although the postwar agricultural crisis could not have been solved through cooperation between the departments, some of its worst effects might well have been mitigated.

[75] C. S. Barrett to C. Bascom Slemp, October 10, 1924, Coolidge Papers, File 227, Box 153; and Michael to Taylor, September 22, 1924, Taylor Papers.

Chapter XI / THE
McNARY-HAUGEN BILL

DESPITE the work of Wallace, his department, and congressional farm representatives, agricultural conditions failed measurably to improve. While the country's economy in general registered a steady rise after the short depression of the early 1920's, the farm sector remained depressed. Prices on agricultural products had fallen to 65 per cent of their prewar purchasing power by the end of 1920 and, with the exception of a few individual commodities such as cotton and wool, had risen only slightly by the middle of 1923. Wheat was a prime example of the problems facing many farmers. In the immediate postwar period it had sold at over $2.50 a bushel, but during 1920 the price declined by more than half. After a modest rise in 1921, wheat offerings turned down again and by the middle of 1923 had reached the prewar price of a dollar a bushel. At the same time, prices of the things farmers had to buy and the taxes they had to pay remained substantially above prewar levels.[1]

Wallace's program of relief and reform, as previously noted, was based upon the belief that the postwar farm crisis would be short-lived. Higher tariffs and emergency credit in particular

[1] *Agricultural Situation* (USDA), August, 1924, p. 20 and October 1924, p. 24; and Henry C. Wallace, *The Wheat Situation: A Report to the President* (USDA, 1923), p. 37.

were to help the agricultural industry through its period of acute depression. Once recovery had begun, government regulation of handlers and processors of farm goods and, in particular, the department's agricultural economics program were to permit producers steadily to improve their position in the economy. But the first step stubbornly refused to materialize, and it became increasingly apparent that the farm legislation of 1921–1923 and the services of the BAE were inadequate to meet the immediate crisis. As a result, Secretary Wallace was forced to conclude that extraordinary measures were necessary if American agriculture were to recover.[2]

The plan Wallace finally settled upon was one formulated by George N. Peek and Hugh S. Johnson. These two men had become friends during the war when both were serving under Bernard Baruch on the War Industries Board. After the armistice, Peek accepted a position as head of the Moline Plow Company and asked Johnson to serve as general counsel of the company. In a precarious financial state at the time Peek assumed its direction, the firm was poorly prepared to weather the postwar agricultural depression. "The implement industry," Peek explained, "is at its lowest ebb . . . [because] the buying power of its customer, the farmer, is paralyzed." Before the industry could hope to recover, he asserted, "it is necessary to restore the farmers' implement dollar." Thus Peek and Johnson devised a scheme designed to aid agriculture, and thereby to improve business conditions in the implement industry and, hopefully, to rescue their own company from bankruptcy.[3]

---

[2] Henry C. Taylor, "Henry C. Wallace and the Farmers' Fight for Equality," p. 3, ms. in Henry C. Taylor Papers, State Historical Society of Wisconsin, Madison.

[3] Address by Peek before the Ohio Implement Dealers' Convention, November 15, 1922, copy in Taylor Papers; and Hugh S. Johnson, *The Blue Eagle from Egg to Earth* (Garden City: Doubleday, Doran, 1935), p. 104. Gilbert C. Fite, the principal historian on Peek and his agricul-

They first presented their plan to a group of businessmen in the fall of 1921. Although the proposal was not seriously received, a somewhat facetious suggestion made an interesting contribution to its history. One of those present thought that the scheme needed a label which would attract the attention of farmers and their representatives. The suggestion appealed to Peek and Johnson, and not long after the meeting they seized upon the phrase "Equality for Agriculture" as representative of the objectives of their plan.[4] The idea expressed by the label was to catch the fancy of agricultural interests and to draw the support of many who only vaguely understood the complicated mechanism of the proposal.

"Equality for Agriculture" became the title of a pamphlet written by Peek and Johnson and published anonymously in early 1922. In it they analyzed conditions contributing to the postwar crisis and concluded that tariffs afforded little benefit to agriculture. The existence of large surpluses in most commodities, according to the pamphlet, rendered farm duties ineffective no matter how high they might be placed. With its control over output and its organization for withholding surplus, industry was able to guard against the price-depressing effects of overproduction. Farmers, on the other hand, had little control over total production and inadequate organization for keeping their goods from flooding the domestic market. "Industry receives, and its prices reflect, the full tariff differential over world price . . . ," the

tural work, maintains that the desire to sell implements only partially explains Peek's interest in farm problems. According to Fite, he was an agrarian at heart and believed that agriculture was basic to the country's economy. See *George N. Peek and the Fight for Farm Parity* (Norman: University of Oklahoma Press, 1954), p. 43.

[4] Alice M. Christensen, "Agricultural Pressure and Government Responses in the United States, 1919–1929," Doctoral dissertation, University of California, Berkeley, 1936, p. 118; and Murray R. Benedict, *Farm Policies of the United States, 1790–1950* (New York: Twentieth Century Fund, 1953), p. 209.

pamphlet explained, but agriculture had to accept prices established in the world market.[5]

It was plain, Peek and Johnson asserted, that farmers operated at a considerable disadvantage. While receiving world prices for their products, they had to pay protected prices for the things they purchased. "If there were no industrial tariff," they reasoned, "he [the farmer] would pay a price fixed by world conditions and his crop would have a fair exchange." As it was, though, the value of agricultural commodities in relation to industrial goods had steadily decreased in the postwar period to the point where the "spread between crop and other prices in this country [is] so great that the farmer . . . is almost completely deprived of buying power." And if a solution to the problem of farm surplus were not found, the process would surely continue.[6]

For Peek and Johnson, however, the solution did not lie in the abandonment of protectionism. Committed to economic nationalism, they felt that the stability of the country's economy depended upon retention of a high tariff. Instead, the solution was to be found in the effective application of the principle of protection to agriculture. This could be accomplished, they explained, "only by diverting surplus from the domestic market," thereby preventing the world price "from establishing domestic price for the entire crop." Through the disposition of excess production abroad, farmers would enjoy a protected market at home and the prices for their products would bear a "fair-exchange" ratio to the prices of other domestic goods.

[5] *Equality for Agriculture* (Moline: H. W. Harrington, 1922), p. 8.

[6] *Ibid.*, pp. 8–11. Peek and Johnson got their idea on the disparity between the prices of agricultural and industrial products from a 1921 publication of the Agriculture Department entitled *Prices of Farm Products*. The bulletin, written by George F. Warren of Cornell, analyzed the movement of agricultural prices in relation to a general price index and concluded that "practically nothing that the farmer sells can be exchanged for the usual quantity of other things." See USDA Bulletin 999, p. 25.

In addition to analyzing the farm crisis Peek and Johnson also presented their scheme, eventually dubbed the "Peek plan," for the establishment of machinery to divert agricultural surplus from the domestic market. "Surplus" was defined as that portion of a given crop which prevented its price from rising to a level representing a fair-exchange ratio. As the core of their plan they recommended creation of a corporation, the sole function of which would be to channel overproduction abroad. Sponsored by the federal government, the corporation was to be empowered to purchase commodities whose prices had fallen below fair-exchange ratios and to sell them outside of the country at prevailing world prices. The fair-exchange price of a given product was to bear the same ratio to current prices as the average ratio of that commodity to the over-all price level for the years 1905 through 1914. Prewar ratios for each agricultural commodity in relation to a general price index were to be computed, and the corporation would endeavor to maintain these by guaranteeing a market for all products at a price representing established ratios.[7]

The operations and losses of the export corporation were to be financed through a "differential loan" collected on the initial sale of agricultural products. Under the Peek plan, the corporation was to estimate the expenses of marketing a given commodity abroad. This sum would then be withheld from the purchase price of the product on a prorated basis, and a special scrip equal to the amount of the assessed loan was to be issued on each sale. The farmer, therefore, would receive the exchange price less the differential loan in cash and the balance in scrip on the sale of a controlled product. At the end of each year the export corporation was to liquidate, and if there were any money left in the differential-loan fund it was to be used to retire the

[7] *Equality for Agriculture*, pp. 19–22.

outstanding scrip at a proportionate rate. Since the scrip was to be negotiable, the farmer could either sell it at the current offering or wait until the year-end retirement.[8]

Anxious to gain acceptance of the plan by the Harding administration, Johnson sent the proof sheets of the soon-to-be-published *Equality for Agriculture* to Hoover in January 1922 and requested his consideration of the proposal. He warned the Commerce secretary that rejection by the Republican party of the "essential principles deduced in the brief will seriously embarrass it in the coming election, if not permanently." [9] Hoover had a member of his department study and report on the scheme, but he was not much impressed with the idea from the beginning.[10] Although he was striving to find foreign outlets for agricultural surplus, the secretary believed that the ultimate solution to the farm problem lay in balancing production to domestic demand, and the Peek plan offered little inducement for such adjustment. In fact, he was later to comment, it would result in further imbalance by stimulating the output of controlled commodities.[11]

[8] *Ibid.*, pp. 20, 31.

[9] H. S. Johnson to Hoover, January 16, 1922, George N. Peek Papers, Western Historical Manuscripts Collection, University of Missouri Library, Columbia; and address by Peek before the Ohio Implement Dealers' Convention.

[10] Memorandum, Louis Domeratsky to Hoover, February 1, 1922, Files of the Bureau of Foreign and Domestic Commerce, National Archives, File 370, Box 1863 and Herbert C. Hoover Papers, Herbert Hoover Presidential Library, West Branch, Iowa, 1-I/2.

[11] "Report to the President's Agricultural Commission," copy in Taylor Papers and Hoover Papers, 1-I/10; *Annual Report of the Secretary of Commerce, 1922*, p. 20; and Hoover to M. L. Dean, March 12, 1924, Hoover Papers, 1-I/2 and 1-I/5. Hoover and Julius Klein later denied that the Commerce Department's policy was for farmers ultimately to withdraw from foreign markets. But Hoover's statement to the commission in 1924 left no doubt in the matter: "Generally the fundamental need is a balancing of agricultural production to our home demand." Julius Klein to George N. Peek, December 8, 1926, Chester C. Davis Papers, Western Historical Manuscripts Collection, University of Missouri Library, Columbia, Box 15.

Hoover had decided that many of the farmers' economic ills could be remedied through greater efficiency in marketing and distribution, and he was disinclined to accept schemes which sought to raise prices without attacking the problem of waste. Like Wallace, moreover, he believed that the crisis would be temporary, and he looked to "economic forces" to bring about long-term adjustment favorable to agriculture. Most significantly, the proposal advanced by Peek and Johnson called for greater government intervention in the economy than the Commerce secretary was ready to sanction.[12] Rather than endorsing the export corporation plan and rallying the administration behind it, as its authors had hoped, Hoover became one of the measure's strongest opponents. Johnson later conceded that it had been a mistake to attempt to interest him in the idea.[13]

Meanwhile, Peek and Johnson tried to place their plan before the National Agricultural Conference of 1922. They felt that the meeting, representing all of the major farm interests in the country, offered an excellent opportunity to gain national exposure of the yet little-known export corporation idea. If well received, furthermore, it might also furnish a lever for applying pressure to gain administration acceptance. But, as noted, they were blocked from introducing their proposal.[14]

Still determined to get a hearing for the export corporation idea, Peek and Johnson remained in Washington after the adjournment of the agricultural conference. Although Wallace had refused to give his permission to place the scheme before the conference, he had not publicly rejected it, and the two authors were confident that once the secretary fully understood their proposal,

[12] Hoover to E. D. Funk, May 31, 1924, Hoover Papers, 1-I/3; *Annual Report of the Secretary of Commerce, 1922*, p. 28; and Herbert C. Hoover, *The Memoirs of Herbert Hoover* (New York: Macmillan, 1952), vol. 2, p. 200.

[13] Johnson, *The Blue Eagle From Egg to Earth*, p. 105.

[14] See Chapter VII.

he would be won over to its support. On the day following the close of the conference, they visited Wallace and requested that he call together a group to consider the Peek plan. The secretary wanted expert opinion on the measure before taking such a step, however, and assigned Henry C. Taylor and George F. Warren, a consultant to the department, to evaluate the proposal and advise him on it. Reporting a few days later, the two economists declared that "some plan of this kind must be added to the tariff idea in order to make the tariff effective in holding up the prices of products." [15] With this assurance, Wallace called a conference for February 13 to investigate the export corporation scheme.

Composed of representatives of agriculture, business, and government, the conference was generally unsympathetic to the Peek plan.[16] Opposition formed around Julius H. Barnes, formerly chairman of the United States Grain Corporation under Hoover's Food Administration, and a man who, according to Wallace and Johnson, represented the Commerce secretary's views on the farm crisis and his attitude toward the Peek plan.[17] Barnes believed that the agricultural problem, if left alone, would work itself out. "The restoration of this fair balance between farm prices and other commodity prices," he maintained, "would be more

[15] Taylor to Wallace, February 3, 1922, Files of the Secretary of Agriculture, National Archives and Taylor Papers.

[16] Attending the conference were Peek; Johnson; Wallace; Henry C. Taylor, head of the Bureau of Agricultural Economics; Julius H. Barnes, former chairman of the United States Grain Corporation; Charles G. Dawes, Director of the Budget, United States Treasury; James R. Howard, president of the American Farm Bureau Federation; Otto Kahn, a New York banker; F. J. Lingham, of the Lockport Milling Co.; George McFadden, representing cotton interests; Gray Silver, Washington representative of the American Farm Bureau Federation; Frederick B. Wells, a grain dealer; Thomas E. Wilson, president of American Institute of Meat Packers.

[17] Russell Lord, *The Wallaces of Iowa* (Boston: Houghton Mifflin, 1947), p. 240; and Johnson, *The Blue Eagle From Egg to Earth*, p. 105.

stable if attained by natural process." The plan in question, on the other hand, would only further disrupt the economy by artificially stimulating the output of products which the corporation might endeavor to support. Through the dumping of farm surplus abroad, furthermore, the scheme would adversely affect world commerce and thus represented a threat to the country's industrial export trade. Even if it were theoretically sound, Barnes insisted, the proposal was not feasible from a practical standpoint, because it would be impossible to establish an index of fair-exchange ratios which represented parity among all commodities. Peek's plan, moreover, was not applicable to perishable products because it required extended periods of storage. And, finally, Barnes pointed out that the creation of a huge bureaucratic organization would be necessary to carry out the export corporation idea, which would not only be very expensive but would invite political interference in the production of the nation's food supply.[18]

Some of the members at the conference took more moderate, if perhaps cautious, attitudes toward the proposal. Otto Kahn, a New York banker representing the financial community, withheld final judgment on the plan, but warned that "too often new schemes have unforeseen consequences." Before taking any action on the Peek plan, he preferred to await the effect of the farm legislation enacted the previous year. "I do not believe," he cautioned, "we should as yet contemplate so thorough a thing as this." Charles G. Dawes, the Director of the Budget, agreed and felt that "where you undertake to institute a new theory that is revolutionary and paternalistic you want to go pretty slow on it." James R. Howard, president of the American Farm Bureau Federation, was concerned about the "restlessness of farmers" and

[18] "Proceedings of a conference called by Henry C. Wallace to consider Equality for Agriculture, February 13, 1922," pp. 48–50, copy in Taylor Papers.

declared that there was need for immediate action to correct the inequity of agricultural prices. While he did not specifically endorse Peek's proposal, the Farm Bureau president suggested that "if we cannot narrow that margin [of prices], I anticipate more unrest and more outbreaks of socialism in the farm sections than we have ever dreamed of." [19]

Peek and Johnson were primarily interested in using the conference as a vehicle for attracting the support of Secretary Wallace. If he agreed to promote their plan, they reasoned that its chances for acceptance by the administration as a whole would be greatly improved. But Wallace remained noncommittal and avoided any direct statements on the Peek plan. Disagreeing with Howard, he felt that there was no "danger of the farmers of this country taking any extreme or radical view," but he conceded that "we should work through every plan that offers any promise even of alleviating the [agricultural] situation. . . ." [20] He failed to make clear, though, whether an export corporation offered such a promise.

If Wallace was reticent to commit himself on the Peek plan, Taylor came out cautiously in favor of it at the meeting. His study of the export corporation idea had convinced him that it offered the possibility of an effective and workable scheme for restoring the farmers' buying power. The measures passed by the previous Congress, the BAE chief believed, were helpful, but they treated only the symptoms of the agricultural crisis without attacking its basic cause. Peek's proposal, on the other hand, was designed to go directly to the source of the problem—the disparity of prices. For this reason, Taylor favored giving the plan a trial to determine if it could be of benefit to American farmers. [21]

[19] *Ibid.*, pp. 54–57, 63.
[20] *Ibid.*, p. 64.
[21] *Ibid.*, pp. 84–85.

The faction which formed under Barnes' leadership prevented any conclusions upon which future action might have been taken, and thus the conference produced no positive results.[22] Still, the plan had received the exposure which Peek and Johnson desired. Wallace was less than enthusiastic, to be sure, but the conference had at least introduced the export corporation scheme to him and had added to his understanding of it. Of perhaps greater importance, the measure found a capable proponent within the Department of Agriculture, and in the future Peek and Johnson were to use Taylor as a channel through which to keep their idea before the secretary. The bureau chief would prove to be a significant influence in bringing about Wallace's eventual support of the Peek plan.

After the special conference, what little interest there had been in the Peek plan seemed to wane. Taking the view that prosperity would soon return, Harding was confident that farmers could be relied upon to make individual adjustments until natural economic laws resolved the crisis. In the meantime, the government should continue to follow the conservative agricultural policy initiated by his administration. A return to normalcy, the President thought, left no room for anything so radical as a government-sponsored export corporation. Although Wallace continued to remain silent on the subject of Peek's proposal, he was reported to be of the opinion that "there were insuperable difficulties in the practical application of the plan." In his 1921 report, he remarked that "better prices for the crops the farmers have to sell and lower prices for the things they have to buy are far more needed than an opportunity to go further in debt." [23] But it was

[22] Address by Peek before the Ohio Implement Dealers' Convention. Lingham, McFadden, Wells, and Wilson supported Barnes' views. "Proceedings of conference," p. 51.

[23] George N. Peek to Theodore M. Knappen, March 3, 1922, Peek Papers; and "Report of the Secretary," USDA *Yearbook of Agriculture*, 1921, p. 15.

clear that Wallace was unwilling in 1922 to go so far as to back the Peek plan in an effort to correct price inequities. The program enacted by Congress, he reiterated, would aid farmers through the period of depression, after which recovery would render stronger measures unnecessary.

When conditions finally forced Wallace to seek stronger measures, he turned naturally to the Peek plan. Since the conference in February 1922, the scheme had been periodically brought to his attention as a result of its authors' persistent campaign to gain acceptance of their proposal. In October 1922 Peek and Johnson published the second edition of *Equality for Agriculture*. This issue, however, was signed and contained only an analysis of the farm crisis and its causes, without a specific plan for bringing about fair-exchange ratios on agricultural products. At the February conference much of the objection had focused on the mechanics of the proposed export corporation rather than on the idea of making tariff protection effective for agriculture. Thus in writing the second edition of the pamphlet, Peek explained to Wallace, "we undertook to get the principles over, leaving a definite plan to be worked out." [24] Peek and Johnson wrote numerous articles, gave talks, corresponded with hundreds of people, and conducted a personal lobby in Washington, all in an effort to stimulate interest in their ideas.[25] They also maintained contact with Taylor, who remained convinced that the two men had "thought out this whole question of the relation of tariff to agriculture more completely than any one else I know of." [26]

[24] George N. Peek to Wallace, January 29, 1924, Peek Papers.

[25] See the Peek Papers for their activities in promotion of the plan from 1922–1924; Hugh Johnson to Hoover, December 23, 1922; and George N. Peek and Hugh Johnson to Frank M. Surface, March 21, 1923, Hoover Papers, 1-I/2 and 1-I/5.

[26] Taylor to George N. Peek, December 14, 1922; Taylor to Peek, December 24, 1922, Peek Papers; and Peek and Johnson to Taylor, March 21, 1923, Taylor Papers.

The Peek plan was less objectionable from Wallace's viewpoint than several other proposals being advanced. Though involving greater government interference in the economy than he would have preferred, the proposal had other features which helped to mollify its undesirable aspects. Because it sought to maintain fair-exchange ratios rather than arbitrarily to set prices, he could satisfy himself that the plan was not a price-fixing measure. Besides, the ultimate purpose was to apply the tariff principle in a way that would provide agriculture with the same kind of protection enjoyed by industry. The differential loan provision also helped Wallace out of a dilemma. Through this assessment, he explained, farmers themselves would pay for the program; the government was only to finance the establishment of the export corporation, after which it would be self-sustaining. Wallace pointed out further that initial capitalization would require about $50,000,000, "the approximate sum which the Government made in the way of profit by its wartime handling of wheat and flour when the price of wheat was arbitrarily controlled and held below the price at which it would have sold without such control." The corporation, furthermore, would be dissolved as soon as "natural economic forces restore normal price ratios." [27] For these reasons, then, Wallace could assure himself that the Peek plan was more acceptable than the "paternalistic" proposals, like Norris' export corporation bill, which envisioned broader and more lasting government involvement in the business of agriculture.

The secretary first announced his endorsement of the Peek plan at a cabinet meeting on September 25, 1923, at which he

[27] "Report of the Secretary," USDA *Yearbook of Agriculture,* 1923, pp. 17–19; United States Congress, House, Committee on Agriculture and Forestry, *Hearings on the McNary-Haugen Export Bill,* 68th Cong., 1st Sess., pp. 114–27; and Henry C. Taylor, "A Farm Economist in Washington," p. 251, ms. in Taylor Papers.

presented a summary of an Agriculture Department report on the situation in the wheat industry. It was incumbent upon the government, Wallace insisted, to help farmers, because the country "cannot hope to have general prosperity until fair price ratios between agriculture, industry, commerce, and labor are restored." The government itself, furthermore, was partly to blame for many of the difficulties which farmers were experiencing. During the war, Wallace recalled, Washington had fixed the price of wheat and effectively controlled the prices of other agricultural products, thus preventing farmers from realizing a full return from their commodities. Appeals by the war administration for more foodstuffs had resulted in increases in cultivated acreage and, eventually, in production which with the drop in postwar demand had brought price-depressing surpluses. Similar appeals to conserve on goods necessary to the war effort had caused decreased per capita consumption of farm products, which by 1923 still had not returned to its prewar average. While the railroads were under federal control, the secretary continued, the government granted substantial wage increases to railway workers, which necessitated rate advances after the lines were returned to private ownership with the end of the war. Because of their dependence on the roads, farmers found that these increases in transportation costs worked a particular hardship on them at a time when the market value of their goods was falling. Wallace further charged that the government's wartime policy of cost-plus industrial contracts and easy credit had encouraged inflation, and he declared that the deflation resulting from the reversal of this policy in the summer of 1920 bore most heavily on agriculture and was a major cause of distorted price ratios. He maintained, therefore, that the government was obligated to assume part of the responsibility for the farm problem and thus to find a remedy for the crisis. "The most hopeful prospect," Wallace concluded, "is the establishment by the Government of an agricultural export commission or corporation with broad powers." [28]

Much to his disappointment, Wallace failed to convince the rest of the administration of the merits of the Peek plan. Had Harding lived, the secretary might well have succeeded in impressing upon him the need for a government-sponsored export corporation. The President, it was true, had initially opposed the idea, but Wallace had always been on friendly terms with him and held considerable influence over the formulation of farm policy in the Harding administration. On the other hand, relations between Coolidge and Wallace were never cordial, and once in the White House Coolidge rarely solicited or accepted his Agriculture secretary's advice. "He [Coolidge] knew not the land beyond the Alleghanies," Taylor complained, and since New England farmers were relatively better off than those farther to the west, "Wallace made little headway in showing him the needs of agriculture. . . . The Coolidge ear was inclined [instead] toward the advisers with whose language he was familiar." [29] The adviser with whose language he was most familiar was Hoover, and the President looked to him rather than Wallace for agricultural advice.

Beyond the lack of communication between the two men, there was a further obstacle in the way of Wallace's demonstrating to the President the need for an export corporation. Unlike Harding, Coolidge was a man of strong convictions, and the Peek plan was contrary to his philosophy of economic individualism. Of the idea that no group should become dependent upon the government for its well-being, the President preached that "wealth comes from industry and from the hard experience of human toil." Coolidge also maintained that the key to national

[28] "Notes of a Cabinet Meeting, September 25, 1923," copy in Taylor Papers; see also *New York Times*, September 26, 1923, p. 1.
[29] Taylor, "A Farm Economist in Washington," pp. 251–52.

progress was economy, and throughout his administration he sought ways to trim the federal budget. And of perhaps greatest importance, he had no interest in unconventional economic ideas.[30] Small wonder, therefore, that the President gave little consideration to the proposal advanced by Peek and Johnson and promoted by Wallace.

Having failed to secure the backing of the rest of the administration, Wallace undertook to stimulate outside interest in the Peek plan. On his quick trip through the West with Harding the previous June and July, Wallace had encountered considerable sentiment among farmers for the export corporation proposal.[31] In October he sent Taylor on a more extensive tour of the wheat states of the Northwest to investigate conditions and to determine the attitude of producers toward the scheme. After numerous conferences with farmers in the region, the BAE chief reported that he had found overwhelming support for the creation of a government export corporation, and he ventured that "nothing short of this will save the agriculture, the banks, and the administration in this whole territory." [32] The enthusiasm registered throughout the spring wheat area undoubtedly encouraged Wallace to redouble his efforts to gain acceptance of the idea.

Taylor had originally planned to visit the winter wheat regions of Kansas and Oklahoma on his way back to Washington, but the trip was cut short. After the BAE chief had been out about

[30] Calvin Coolidge, *The Autobiography of Calvin Coolidge* (New York: Cosmopolitan Book Corporation, 1929), p. 182; and Claude M. Fuess, *Calvin Coolidge: The Man from Vermont* (Boston: Little, Brown, 1940), p. 384.

[31] United States Congress, Senate, Committee on Agriculture and Forestry, *Hearings on the Purchase and Sale of Farm Products,* 68th Cong., 1st Sess., p. 357.

[32] Telegram, Taylor to Wallace, October 17, 1923, Taylor Papers. These papers contain several telegrams sent to the department by Taylor, all reporting strong support for the Peek plan. He visited North Dakota, Montana, Idaho, Washington, and Oregon. See also Taylor, "A Farm Economist in Washington," pp. 286–88.

two weeks, Coolidge ordered Wallace to recall him. This action was purportedly taken at the request of future Secretary of Agriculture William M. Jardine, president of Kansas State College, who wanted no one politicking for the Peek plan in his state.[33] Jardine's attitude, as he was to announce later, was that Kansas farmers should be left alone: "We will work ourselves out of the hole we are in." [34] Coolidge, of course, could applaud this kind of individual determination and apparently complied with the request without hesitation.

Upon returning to Washington, Taylor prepared a report to the President on his findings but was never given an opportunity to submit it. No doubt Coolidge's lack of interest in the report was due to the fact that he had sent a commission of his own into the Northwest, which was touring the area at the same time as Taylor. Headed by Eugene Meyer and Frank W. Mondell of the War Finance Corporation, the Coolidge group was clearly intended to counteract the influence of the Agriculture Department and to stifle any movement for the export corporation. Not consulted on the matter, Wallace was reportedly very unhappy with the President's action.[35] After the project was publicly announced, Coolidge requested that the secretary name a representative of the Agriculture Department to the commission. Meyer wanted Taylor or his assistant, Lloyd S. Tenny, to accompany the group, but Wallace refused to appoint a top official of his department and assigned instead an underling, H. S. Yohe. Largely ignored by Meyer and Mondell, Yohe was given small part in the work of the commission. When the tour was about half com-

---

[33] Henry C. Taylor and Anne D. Taylor, *The Story of Agricultural Economics, 1840–1932* (Ames: Iowa State College Press, 1952), p. 595.

[34] *Washington Post*, May 15, 1924, p. 1. In a long letter to Hoover, Jardine outlined his objections to the McNary-Haugen bill. William M. Jardine to Hoover, April 14, 1924, Hoover Papers, 1-I/8.

[35] Lloyd Tenny to Taylor, October 16, 1923, Taylor Papers; and Taylor, "Henry C. Wallace and the Farmers' Fight for Equality," p. 4.

pleted, the project leaders suggested that it was not necessary for Yohe to finish the trip, and he returned to Washington.[36] The existence of the second commission and the treatment of Agriculture's representative were indicative of the gulf which had grown between Wallace and his department on the one hand, and Coolidge and the rest of the administration on the other.

Before leaving, Meyer held conferences in New York with several grain marketing representatives, one of whom was Hoover's former subordinate, Julius H. Barnes. The country's grain dealers had voiced strong disapproval of the Peek plan from the beginning, and the meetings were presumably to arm Meyer with arguments against the scheme which he could present to wheat farmers on his tour.[37] The commission's first stop was Chicago, where the group conferred with Aaron Sapiro, a leader in the cooperative marketing movement. At this meeting, plans were drawn for the formation of a nationwide grain marketing association, which was to be the administration's alternative to an export corporation for wheat farmers. The organization was, of course, to be voluntary, and the government's role would be advisory, with perhaps some aid in the form of loans through the War Finance Corporation.[38]

The Meyer-Mondell commission met with small success in the Northwest. By the group's own admission, it encountered the same strong sentiment for the Peek plan that Taylor reported.[39] Interestingly, some of the marketing associations were among the

[36] Lloyd Tenny to Taylor, October 16, 1923, Taylor Papers; H. S. Yohe to Wallace, October 9, 1923; and H. S. Yohe to Wallace, October 16, 1923, Files of the Secretary of Agriculture.

[37] New York Times, October 2, 1923, I, p. 12.

[38] H. S. Yohe to Wallace, October 9, 1923, Files of the Secretary of Agriculture; Lloyd Tenny to Taylor, October 16, 1923, Taylor Papers; and New York Times, October 8, 1923, p. 26.

[39] Eugene Meyer and Frank W. Mondell to Coolidge, November 5, 1923, Calvin Coolidge Papers, Manuscript Division, Library of Congress, File 157, Box 127.

strongest in opposition to the administration's suggestion that co-operative activity offered the best solution to the farmers' problems. Walter J. Robinson, president of the Wheat Growers' Association, asserted that "farmers are not willing to undertake co-operation on a large scale without first having assurance that the surplus would be taken care of in such a way that an arbitrary price can be placed on wheat for domestic consumption." A Spokane newspaper echoed Robinson's remarks when it wrote that marketing associations of the Pacific Northwest opposed Meyer's proposal because "no organization, cooperative or otherwise, can fix the price [of wheat] so long as there is an exportable surplus of nearly 200,000,000 bushels." [40] Not only were agricultural interests in the area reluctant to accept the commission's proposition, but they were also suspicious of the motives of the administration. Robinson reported that it was "almost universal opinion that Mr. Meyer was sent West to induce the farmers to get their minds on co-operation in order that they might forget government aid." A grain dealer, surprisingly, referred to the tour as a "political junket" and indicated that producers were "extremely resentful toward the administration for sending a commission to investigate the condition of the Northwest wheat farmer, which arrived with preconceived ideas as to the proper remedy to be applied." Chester C. Davis, Montana agricultural commissioner and soon to become a key figure in the movement for the Peek plan, thought that the "ambassadors from the War Finance Corporation . . . did not come to this State to seek information or to try to learn anything," but only to solicit statements "corroborative of the point of view they sought to convey." [41]

[40] Walter J. Robinson to Taylor, October 26, 1923, Taylor Papers; and *Spokesman Review,* October 26, 1923, p. 1.

[41] Walter J. Robinson to Taylor, October 26, 1923; Frederick B. Wells to Taylor, October 24, 1923; T. S. Hedges to Taylor, November 19, 1923; and Chester C. Davis to Taylor, October 23, 1923, Taylor Papers.

Meyer and Mondell devoted a good deal of their report to the President to attacking the Peek plan. With more than a little logic, they maintained that it would be impossible to determine what portion of a given crop represented a "surplus." And still on firm ground, they pointed out that the mechanism of the proposed corporation would be extremely complicated and most difficult to put into operation. But, apparently forgetting that the same argument might well be used against the high tariff policy of the administration they were representing, Meyer and Mondell went on to charge that the dumping of surplus abroad would bring retaliatory measures from other agricultural exporting countries. The two men also reported that they had found strong approval for the expansion of cooperative marketing as a substitute for the export corporation, but without indicating that it was primarily among bankers and businessmen rather than farmers. The report concluded by recommending voluntary production adjustment and an improved system of marketing associations as the most effective solutions to the agricultural crisis.[42]

If Wallace had any hope of rallying the administration behind the Peek plan, the activities and report of the Meyer-Mondell commission clearly demonstrated that it was baseless. Thus the secretary set out on a course divergent to that of Coolidge and the other members of the cabinet. While attending the convention of the American Association of Land Grant Colleges and Universities, he was invited to address the Chicago Association of Commerce. In that speech, delivered on November 14, 1923, Wallace gave his first public endorsement of Peek's proposal. Contending that the agricultural situation had reached a point where it demanded a more aggressive policy than was currently being pursued, he suggested that the creation of a government-sponsored export corporation was the action most likely to bring

[42] Eugene Meyer and Frank W. Mondell to Coolidge, November 5, 1923, Coolidge Papers, File 157, Box 127.

the rapid relief which farmers had to have. In his published report to the President two weeks later on the wheat situation, Wallace stated that "an export corporation to aid in the disposition of . . . surplus is worthy of the most careful consideration." [43]

Meanwhile, Wallace had assigned the task of drafting a tentative bill incorporating the Peek plan to Charles J. Brand, a member of the Agriculture Department. Brand traveled to Chicago to confer with Peek and Johnson on the project. His completed bill was subsequently reworked by the drafting sections of the Senate and House under the sponsorship of two congressmen from agricultural states, Charles L. McNary of Oregon and Gilbert N. Haugen of Iowa.[44] And in January 1924 the first of several McNary-Haugen bills was introduced into Congress.

The bill was essentially the same as the plan presented in the first issue of *Equality for Agriculture*. It called for the creation of an export commission consisting of the Secretaries of Agriculture, Commerce, and Treasury, the Chairman of the Tariff Commission, and three directors to be appointed by the President with the consent of the Senate. The commission was to establish fair-exchange ratios based on averages for the years 1905 through 1914 for an enumerated list of agricultural and processed products, and to determine when domestic prices fell below ratio prices. If the domestic price of a given product were too low, the President was then authorized to declare a special emergency and to empower a government corporation to enter the market and purchase that commodity at the ratio price.[45] Capitalized at $200,000,000 subscribed by the federal govern-

[43] *New York Times,* November 15, 1923, p. 30; and Wallace, *The Wheat Situation,* pp. 74–75.

[44] *Hearings on the McNary-Haugen Export Bill,* p. 43; and *Hearings on the Purchase and Sale of Farm Products,* p. 578.

[45] The list included wheat, flour, corn, raw cotton, wool, cattle, sheep, swine, and all products of cattle, sheep, and swine.

ment, the corporation would be able to sell on the foreign market at the world price or to store the surplus for future resale on the domestic market at no less than the ratio price. Farmers could sell to the corporation or to private dealers, but in either case a differential loan, or "equalization fee" as it was called in the bill, would be collected and deposited with the corporation. Once the domestic price of a controlled commodity reached the ratio price, the corporation would withdraw from the market and cease purchase of that product until the price again fell below the fair-exchange ratio. The bill also provided for flexible tariffs on all enumerated items equal to the difference between the world and ratio prices, which would be automatically adjusted as prices fluctuated.[46]

During the hearings on the bill, lines of support and opposition were sharply drawn. Processors and handlers of farm products were unanimously against the measure, principally because it proposed further government interference in the private sector of the economy. For much the same reason, the American Bankers' Association, representing primarily eastern bankers, came out early in opposition to the export corporation idea. Bernard Baruch, who had endorsed the Peek plan from the beginning, and Otto Kahn, who gave qualified support to the bill, were the major exceptions among eastern financial interests. Western bankers servicing rural areas, on the other hand, generally backed the measure. The National Chamber of Commerce, with Julius Barnes as its new president, also applied its influence to secure defeat of the bill.[47]

[46] See *Hearings on the McNary-Haugen Export Bill,* pp. 19–28 for copy of the bill.

[47] See *ibid.;* Walter J. Robinson to Taylor, November 12, 1923, Taylor Papers; Wallace to George N. Peek, April 30, 1924, Peek Papers; Otto Kahn to Wallace, March 8, 1924; *The Situation in Washington* (Sears, 1924), copy in files of the Secretary of Agriculture; Charles N. Herreid to Hoover, April 2, 1924, Hoover Papers, 1-I/9; Julius H. Barnes to

Before the hearings closed, the House Committee on Agriculture and Forestry had received over 10,000 endorsements of the McNary-Haugen bill. Although many of those citations came from agricultural organizations, their support was neither unanimous nor enthusiastic. Backing from the Farm Bureau was slow and qualified by a recommendation to amend the bill to insure protection of cooperatives. The National Farmers' Union preferred a reintroduced Norris bill, but several of its state divisions came out in favor of the McNary-Haugen bill. Similarly, the executive committee of the Grange spoke out against the legislation, while local organizations petitioned Congress in its behalf. Cooperative marketing associations were also divided, with some, such as the American Wheat Growers' Association, strongly in support and others, like the American Dairy Federation and the Oklahoma Wheat Growers' Association, in opposition. Several middle-western agricultural journals, most notably *Wallaces' Farmer* and the *Prairie Farmer*, editorialized for the measure. Alignment among farm interests was generally along sectional lines, with the spring wheat regions of the Northwest most actively in favor of the bill.[48] In the winter wheat, corn, and dairy areas, where conditions were better, sentiment was slower to develop and never as enthusiastic, and in the South

Coolidge, March 8, 1924, Coolidge Papers, File 227C, Box 156; *New York Times*, May 22, 1924, p. 19, September 27, 1923, p. 9, and September 28, 1923, p. 1.

[48] Louis G. Michael to Wallace, January 24, 1924, Files of the Secretary of Agriculture; Charles E. Hearst to Albert B. Cummins, May 3, 1924, Albert B. Cummins Papers, Iowa State Department of History and Archives, Des Moines, Box 27; Taylor to Dwight R. Cresap, October 22, 1923, Taylor Papers; John Manley to George N. Peek, March 6, 1924; Circular letters by George C. Jewett, May 16, 1924 and May 17, 1924, Peek Papers; "Resolutions Adopted and Passed by the Oklahoma Wheat Growers' Association, February 21, 1924," copy in Hoover Papers, 1-I/7; *Wallaces' Farmer* 48 (December 28, 1923): 1754 and 49 (March 14, 1924): 240; and *Prairie Farmer*, February 23, 1924, p. 4 and March 15, 1924, p. 3.

sharp opposition arose due to traditional attitudes toward protectionism and an upward trend in cotton prices.

While representatives of various interest groups laid their arguments before the Senate and House Committees on Agriculture and Forestry, the administration was busy formulating a program to divert attention from the McNary-Haugen bill. Although Coolidge never publicly committed himself on the measure, it was common knowledge that he opposed its enactment.[49] In his address to the opening of the Sixty-eighth Congress in December 1923, he indirectly attacked the proposal. "No complicated scheme of relief, no plan for Government fixing of prices, no resort to the Public Treasury," the President lectured, "will be of any permanent value in establishing agriculture." Coolidge believed that "simple and direct methods put into operation by the farmer himself are the only real sources for restoration." Following closely the recommendations of the Meyer-Mondell commission, he suggested production adjustment and cooperation as the best ways to meet the problems of farmers. Included as part of an agricultural adjustment program, the President explained, should be not only the reduction of acreage in surplus crops but diversification as well. The government's role in both of these projects was to be as a source of loans, with development by farmers themselves on a voluntary basis.[50]

Early the following year Peter Norbeck of South Dakota and Olger B. Burtness of North Dakota introduced a bill into Congress providing for federal assistance in the diversification of agriculture. Based upon a plan formulated by John L. Coulter, president of North Dakota State Agricultural College, the measure provided for the creation of a Federal Agricultural Diversifi-

[49] *New York Times,* March 5, 1924, p. 10 and May 6, 1924, p. 3; C. Bascom Slemp to M. L. Bowman, February 23, 1924, Coolidge Papers, File 227, Box 153; and Albert B. Cummins to Don F. Berry, March 31, 1924, Cummins Papers, Box 27.

[50] *Congressional Record,* 68th Cong., 1st Sess., p. 100.

cation Commission, consisting of the secretaries of Commerce, Agriculture, and the Treasury. With an appropriation of $50,-000,000, the commission was to make loans to farmers for the purpose of diversifying their operations.[51] Coolidge, at whose instigation the bill had been introduced, immediately endorsed the Norbeck-Burtness measure as an alternative to the McNary-Haugen bill. Also in support of the legislation, Hoover revealed that "some change in the basis of agriculture by diversification and a larger direct production of the family needs [is] to my mind the ultimate solution." [52]

Secretary Wallace, on the other hand, conceding that diversification held out the possibility of long-range improvement in some regions, insisted that it would not provide the immediate solution to price disparity which farmers needed. He also pointed out that the proposal was impractical for many producers who were limited by climatic and soil conditions and by experience to a single type of operation. But, apparently fearful that outright opposition might be misinterpreted, the secretary remained silent on the Norbeck-Burtness bill.[53] The proposal stimulated little interest among farmers, however, mainly because there were so few realistic possibilities with which to diversify. According to one agricultural representative, anything producers might substitute for wheat was "about equally unprofitable." Defeat of the bill in March 1924, Chester C. Davis observed, "failed to cause even a

[51] Taylor to John M. Davis, February 15, 1924, Taylor Papers; and Christensen, "Agricultural Pressure and Government Responses in the United States," pp. 141–42.

[52] Coolidge to C. T. Jaffray, undated, Coolidge Papers, File 227D, Box 157; Hoover to C. N. Herreid, December 18, 1923; Hoover to Julius H. Barnes, November 30, 1923; Hoover to Olger B. Burtness, January 17, 1924; and Hoover to Frederick E. Murphy, April 3, 1924, Hoover Papers, 1-I/9.

[53] Wallace to George W. Young, June 11, 1923; Wallace to A. W. Lindsay, November 17, 1923, Files of the Secretary of Agriculture; Wallace to M. W. Heptonstall, May 23, 1921, Files of the Bureau of Agricultural Economics, National Archives.

ripple of disappointment among the farmers. Their eyes are on the McNary-Haugen bill." [54]

In another effort to meet growing demand in the Northwest for stronger government action against the farm crisis, Coolidge called an agricultural conference to study the financial situation in this area. The President consulted Hoover on the makeup of the conference, and the Commerce secretary responded by recommending delegates drawn largely from the business-financial world and not at all from agriculture. Although Hoover belatedly requested Wallace's comments on his list, the Agriculture secretary participated in neither the selection of the delegates nor in the planning of the program. The final composition of the meeting followed closely Hoover's suggestions, except that officers of the leading farm organizations were also included. Invitations were sent out from the President's office rather than by the Agriculture Department, as had been the case in the 1922 conference.[55] Despite the fact that Wallace said nothing publicly, he must have resented being by-passed in the preparation and calling of this meeting.

Convening on February 4, 1924, for a one-day session, the conference was obviously intended merely to sanction the administration's farm proposals. As chairman of the meeting, Hoover completely dominated the proceedings, guided the discussion, and discouraged unwanted resolutions. Wallace, after a short and harmless speech at the opening of the conference, took no further part. At Coolidge's urging, the group endorsed the Norbeck-Burtness bill, called for higher duties on wheat, and

[54] George C. Jewett to Coolidge, April 8, 1924, Coolidge Papers, File 277C, Box 156; and Davis to George N. Peek, March 15, 1924, Peek Papers.

[55] Christian A. Herter to C. Bascom Slemp, January 28, 1924; C. B. Slemp to R. B. Bissell, January 29, 1924, Coolidge Papers, File 227D, Box 157; and Hoover to Wallace, January 28, 1924, Hoover Papers, 1-I/2. For the list of delegates see Coolidge Papers, File 227D, Box 157.

adopted a plan for the expansion of rural credit facilities through private sources.[56] Although the meeting blocked a motion by farm representatives to recommend enactment of the McNary-Haugen bill, the agricultural faction met with Wallace after adjournment of the regular session and drew up a minority report in support of the measure.[57]

Soon after the close of the conference, a group of financiers followed up Coolidge's suggestion for a private farm loan agency by forming the Agricultural Credit Corporation. Subscribed entirely by private investors, the corporation made loans to rural banks on agricultural paper as well as directly to farmers. With the defeat of the Norbeck-Burtness bill, the main activity of the agency was to advance credit for the purpose of diversification, principally in the spring wheat region. In recommending this move, Coolidge declared that he knew of "no more effective service that could be rendered to the agricultural interests of the central northwest."[58] John L. Coulter, author of the plan on which the Norbeck-Burtness bill had been based, revealed the administration's objective when he noted with satisfaction that farmers were following the diversification program of the Agricultural Credit Corporation and "ignoring the agitation being carried on by the politicians who are promising to provide some artificial stimulus to the wheat market thru the McNary-Haugen bill or otherwise."[59] Though the total amount extended in loans was

[56] "Minutes of President's Conference On Northwestern Agriculture and Finance, February 4, 1924," copy in Hoover Papers, 1-I/9; and Coolidge to C. T. Jaffray, undated, Coolidge Papers, File 227D, Box 157.

[57] "Minutes of Meetings of Delegates Directly Representing Agriculture in the Financial and Agricultural Conference Called by President Coolidge on February 4," copy in Taylor Papers.

[58] Eugene Meyer, Jr. to Coolidge, May 12, 1924, Coolidge Papers, File 157, Box 127; Coolidge to C. T. Jaffray, undated, *ibid.*, File 227D, Box 157; and Hoover to Frederick E. Murphy, April 3, 1924, Hoover Papers, 1-I/9.

[59] John L. Coulter to Eugene Meyer, Jr., May 23, 1924, Coolidge Papers, File 157, Box 127.

small and the number of banks and individual farmers aided negligible, the administration could take credit for creation of the corporation and thus for providing assistance to farmers.[60]

Coolidge also had the loan-making period of the War Finance Corporation, due to expire on July 31, extended to the end of 1924. In February the corporation established a branch in Sioux Falls, South Dakota, for the purpose of aiding rural banks in the Northwest until the farm crisis had passed. Like the Agricultural Credit Corporation, the War Finance Corporation turned its attention primarily toward the promotion of diversification in the spring wheat region.[61]

As the price of wheat continued to decline in 1923, producers began calling for further increases in the tariff to block Canadian imports. They appealed to the President to exercise his power under an elastic provision of the Fordney-McCumber Tariff to raise the duty on wheat 50 percent upon recommendation of the Tariff Commission. Coolidge announced in late September 1923 that, while the existing tariff was of benefit to producers, he opposed any further rate increases at the time. In the midst of mounting sentiment for the McNary-Haugen bill, however, he decided four months later that an advance in the tariff on wheat was necessary. The President called for a favorable ruling by the Tariff Commission, which had been studying the question since the previous November, and induced the February agricultural conference to endorse such a move. Finally, in early March 1924, the Tariff Commission recommended that duties on wheat

---

[60] In its first year, the Agricultural Credit Association helped 226 banks and loaned an additional $600,000 to some 1300 individual farmers. Fiscal Report, December 31, 1924, copy in Hoover Papers, 1-I/9. For Coolidge's comments on the association, see the *Country Gentleman* 89 (May 4, 1924): 4.

[61] Eugene Meyer, Jr. to Coolidge, March 12, 1924, Coolidge Papers, File 157, Box 127; and *New York Times,* February 3, 1924, II, p. 4 and April 6, 1924, p. 16.

be raised from 30 to 42 cents a bushel, and Coolidge quickly ordered the increase.[62]

The principal measure advanced by the President as an alternative to the McNary-Haugen bill remained federal aid to cooperative marketing. Chief architect of the plan which the administration endorsed and attempted to force through Congress in 1924 was Secretary Hoover. He believed that a system of orderly marketing through nationwide cooperative associations offered the most feasible solution to the postwar crisis. Convinced that inefficiency in the distribution of products was a major cause of agricultural problems, he envisioned a vast network of farmer-controlled agencies to regularize the flow of commodities to market. Once a balance had been struck between supply and demand and the delivery of goods had been improved, he reasoned, prices could be stabilized behind tariff walls at a level representing a fair return to producers. While Hoover expected farmers to make the necessary production adjustments, he felt that the development of a sound system of marketing associations required federal assistance and supervision. Formation of cooperatives was to be on a voluntary basis, however, and their operation largely in the hands of producers. Such a program of self-help naturally appealed to Coolidge, for it held out the possibility of relieving the farm depression with a minimum of government intervention.[63]

After an extensive investigation of cooperative marketing, the Commerce Department drew up a bill embodying Hoover's

[62] George Severance to Taylor, November 8, 1923, Taylor Papers; Charles J. Brand to Wallace, November 15, 1923, Files of the Secretary of Agriculture; *New York Times,* September 26, 1923, p. 4 and February 26, 1924, p. 28; and *Wheat and Wheat Products: Report of the United States Tariff Commission* (1924).

[63] Memorandum, E. G. Montgomery to Hoover, December 15, 1922, Hoover Papers, 1-I/3; Hoover *Memoirs,* vol. 2, p. 110; and James H. Shideler, "Herbert Hoover and the Federal Farm Board Project, 1921–1925," *Mississippi Valley Historical Review* 42 (March, 1956): 712–16.

ideas, which the secretary sent to Michigan Representative Arthur B. Williams in March 1924 with a note expressing the desire that he would "father the matter." "It is a bolder attack on the whole problem than ever hitherto attempted," Hoover told him, "for I have felt that we would not secure any strong and permanent development of cooperatives until we lay the foundation for embracing the whole distribution problem in a complete system." [64] Early the following month Williams introduced a bill drawn from the department's draft, and Arthur Capper of Kansas sponsored a companion measure in the Senate. The Capper-Williams bill provided for the formation of a Federal Marketing Board to promote the development of cooperative marketing and to license and regulate the associations. Working under the board, separate commodity divisions were to advise the organizations and to enforce standard operational procedures. An appropriation of $10,000,000 was to be used for loans to cooperatives to aid in their formation and marketing activities.[65]

Wallace set himself firmly against the Capper-Williams bill. In response to Coolidge's request for his views on the measure, he expressed the belief that for the government actively to promote cooperatives would in the long run prove harmful to the movement, for the success of such enterprises "ultimately depends upon the confidence, loyalty and wholehearted voluntary support of the individual member." Wallace favored the policy of the Agriculture Department, which limited its services to advising without engaging in organizational or managerial activities. As he had several times before, the secretary insisted that the federal government "should help the farmers market their crops just as it helps them to produce crops, not by doing the work but by

[64] Hoover to Leonidas Y. Redwine, August 13, 1923; and Hoover to Arthur B. Williams, March 29, 1924, Hoover Papers, 1-I/3.
[65] Shideler, "Herbert Hoover and the Federal Farm Board Project," p. 719.

supplying information which the farmers can not get for themselves." In Wallace's opinion, furthermore, the Capper-Williams proposal was inadequate to meet the needs of the depressed farming industry, and "for the administration to put forward or support this or any similar bill with the suggestion that it will relieve the agricultural situation . . . would not be accepted by farmers generally as offering any substantial measure of relief from their present problems." [66] Despite Wallace's protestations against increased federal interference, however, he could not have been overly concerned, for the Peek plan went considerably further in that direction than Hoover's cooperative marketing plan. The secretary's major objections to the Capper-Williams bill were clearly ones which he failed to mention. That it represented yet another instance of encroachment by the Commerce Department upon Agriculture's domain and that it was being advanced, at least in part, to undermine the movement for the McNary-Haugen bill were undoubtedly more important factors in shaping his attitude toward the measure than the reasons he gave.

The Capper-Williams bill elicited little enthusiasm among agricultural interests. Weary of programs requiring long-term development, farmers were looking for measures providing immediate relief. Many agreed with one of their representatives who wrote to Williams that "the present basic cause of the farmer's ills is over-production, i.e., production far in excess of possible consumption. Cooperative marketing will not cure that ill." O. E. Bradfute, the recently elected president of the American Farm Bureau Federation, advised Coolidge that "farmers are very restless under present conditions and feel that a greater effort

[66] Coolidge to Wallace, March 29, 1924 and attached note of Wallace's views, Coolidge Papers, File 227B, Box 155; Wallace to Coolidge, April 8, 1924, Hoover Papers, 1-I/5; and "Report of the Secretary," USDA *Yearbook of Agriculture,* 1924, p. 44.

should be made to at the present time find a plan to meet the emergency in which we find ourselves." Convinced that the McNary-Haugen bill offered the best solution to the agricultural crisis, Bradfute suggested that farmers "would be very sorry to have that sidetracked for any proposal that might take several years before it could properly function." [67] Other than a sympathetic hearing before the executive committee of the American Farm Bureau Federation and the endorsement of future Agriculture Secretary William M. Jardine and some cooperative leaders, the Capper-Williams bill attracted little support outside of the administration.[68] Due partly to lack of interest and partly to the fact that the bill was introduced late, the first session of the Sixty-eighth Congress failed to consider the measure.

If it accomplished nothing else, therefore, the McNary-Haugen bill forced the administration to formulate and advance agricultural proposals of its own. When Coolidge moved into the Presidency after Harding's death, he had little to offer farmers in the way of a program. As late as his first address to Congress in December 1923, moreover, he could talk only vaguely and in general terms about diversification and cooperative marketing. Referring to the President's message, one wag remarked that the part he liked best was the section in which Coolidge told farmers to go to hell.[69] But rising interest in the McNary-Haugen bill sparked the administration into action and compelled it to work out and take a stand on specific proposals. Farmers might not have been happy with Coolidge's program, but at least they now knew where he stood.

Meanwhile, Wallace and the Agriculture Department were not

[67] A. V. Chaney to Arthur B. Williams, June 2, 1924, Taylor Papers; and O. E. Bradfute to Coolidge, April 15, 1924, Hoover Papers, 1-I/5.

[68] William M. Jardine to Coolidge, April 7, 1924, Hoover Papers, 1-I/5; and Shideler, "Herbert Hoover and the Federal Farm Board Project," p. 718.

[69] Fite, George N. Peek, p. 78.

idle. The secretary ceased publicly campaigning for the Peek plan, but he and his staff worked quietly in the background attempting to gain support for the McNary-Haugen bill. Thus there arose a curious situation in which one department of the executive branch was engaged in activities conflicting sharply with those of the rest of the administration.

Wallace gave Charles J. Brand, author of the department's tentative bill, the job of proselytizing among congressmen for the McNary-Haugen measure. Brand was in frequent contact with Peek, who supplied at his own expense a vast amount of printed material explaining the export corporation idea and urging its adoption. Working closely together, the two men distributed this information, organized supporters of the bill, and applied pressure on Congress and the administration. In addition, Brand presented a strong case for the McNary-Haugen bill at the hearings before the House Committee on Agriculture and Forestry.[70] Besides conducting its own lobby under Brand's direction, the Agriculture Department cooperated with the Interstate Export League, an affiliation of *ad hoc* organizations formed in 17 states to promote passage of the bill. Wallace also engaged Chester C. Davis, who was in Washington representing the Montana Export League, as an unofficial liaison between the Agriculture Department and Congress.[71]

Secretary Wallace himself actively promoted the McNary-Haugen bill. Peek furnished him with copies of *Equality for Agriculture* and other literature, which the secretary distributed among friends on Capitol Hill and various groups interested in the mea-

[70] C. J. Brand to George N. Peek, March 7, 1924; Brand to Peek, March 8, 1924; Brand to Peek, April 29, 1924, Peek Papers; and *Hearings on the McNary-Haugen Export Bill*, pp. 539–46, 602–5.

[71] *Hearings on the Purchase and Sale of Farm Products*, p. 388; Wallace to S. H. Thompson, February 20, 1924, Files of the Secretary of Agriculture; and Chester C. Davis to Taylor, November 19, 1924, Taylor Papers. At the bottom of its official stationery, the Interstate Export League listed Wallace as a supporter of the bill.

sure. Appearing before both the Senate and House Committees on Agriculture and Forestry, he urged favorable action on the legislation. After the bill had been reported out of committee, he appealed to the Iowa delegation in the House for solid support. Using a favorite tactic, Wallace also sent out hundreds of letters urging friends of the measure to contact their congressmen.[72]

Both committees reported the McNary-Haugen bill favorably. Because of the tariff feature, the Senate delayed consideration of the legislation until the House, in which revenue measures have to originate, had acted upon it.[73] Although the House conducted an extended debate, it added little to an understanding of the bill and had practically no effect on the ultimate voting. Before the McNary-Haugen bill reached the House floor in May 1924, the export corporation plan had been discussed and analyzed for two years and lines of support and opposition had been clearly drawn. On the floor proponents and opponents merely rehearsed old arguments, failing measurably to influence the attitudes of their congressional colleagues. The McNary-Haugen bill had been marked as a sectional measure by the time the House came to consider it, and the debate merely reinforced its sectional character.

Opponents of the legislation charged that it was irresponsible

[72] Wallace to Peek, February 5, 1924, Peek Papers; *Hearings on the McNary-Haugen Export Bill,* pp. 114–27; *Hearings on the Purchase and Sale of Farm Products,* pp. 347–76; Wallace to John D. Jones, Jr., January 25, 1924; Wallace to F. F. Frazee, April 21, 1924; Wallace to A. J. Granger, April 2, 1924, Files of the Secretary of Agriculture; and Cyrenus Cole, *I Remember, I Remember* (Iowa City: State Historical Society, 1936), p. 456.

[73] There was a move by Senator Peter Norbeck of South Dakota to attach the McNary-Haugen bill to the tax bill already enacted by the House. But McNary preferred to await House action and had his measure indefinitely postponed. Peter Norbeck to W. H. King, May 7, 1924; Norbeck to King, May 19, 1924, Peter Norbeck Letters, Western Historical Manuscripts Collection, University of Missouri Library, Columbia, Box 1; and *Congressional Record,* 68th Cong., 1st Sess., p. 6760.

and radical. James B. Aswell of Louisiana, the most outspoken critic of the bill, described it as "socialistic, . . . [a] dangerously vicious form of paternalism, . . . [which would] place our Government squarely in the class of the bolshevistic Government of Russia. . . ." Maintaining that the measure was highly inflationary, Wisconsin Representative Edward Voigt feared that it would result in a "pyramid of prices" and thus work hardship on the country's consumers. Other objections leveled against the bill were that it was unworkable and a price-fixing scheme, that it would compound the problem of overproduction and interfere with the normal functioning of the economy, and that it would draw retaliation from other agricultural exporting countries.

Led by Haugen and J. N. Tincher of Kansas, backers countered with equal vehemence that other groups had benefited from such federal assistance as the tariff, Adamson Law, Transportation Act, and various subsidies, and that it was only fair for producers of food and fiber to receive like consideration. Farm problems having resulted at least in part from wartime policies, they further contended, the government was obligated to do something to help rescue the agricultural industry. Since the McNary-Haugen bill provided the most logical and comprehensive approach to the crisis yet offered, Tincher thought that Congress, "in simple justice to the American farmer," ought to adopt the legislation. After a month of sharp debate, the measure was rejected on a rather lopsided tally of 223 to 154. As expected, the vote was largely along sectional lines, with representatives from the Middle and Far West heavily in favor and those from the East and South combining to defeat the bill.[74]

The defeat of the McNary-Haugen bill was due largely to two factors: active administration opposition and the failure of agricultural interests to unite behind it. While the administration was

[74] For debate and vote on the bill see *Congressional Record,* 68th Cong., 1st Sess., pp. 9021–38, 9196–239, 10041–67, 10340–41.

unsuccessful in attempts to institute its own program of federal aid for diversification and cooperative marketing, its influence was a factor in the rejection of the export corporation scheme. During the House debate, opponents revealed that Coolidge had privately indicated he would veto the bill if passed by Congress.[75] On top of the administration's well-known attitude toward the scheme, this prospect undoubtedly deterred some congressmen, who might not have felt strongly one way or the other, from voting for it. More important in the defeat of the bill, however, was division among agricultural groups. The progressive bloc had kept alive George W. Norris' proposal for a government export corporation, which competed with the McNary-Haugen bill for the support of farm representatives. In early 1924 the Nebraska senator reintroduced his bill, and the Senate Committee on Agriculture and Forestry held hearings on both export corporation measures simultaneously. Included in the report on the McNary-Haugen bill was a minority report favoring the Norris bill. The committee eventually sent Norris' measure to the Senate, and, although not acted upon, it distracted attention from the McNary-Haugen bill.[76] In addition to the halfhearted support of major agricultural organizations, furthermore, lack of interest in either proposal among a sizable number of farm repre-

[75] *Ibid.*, p. 9928. Gilbert N. Haugen, the sponsor of the House bill, indicated as early as March that administration opposition might cause defeat of the bill. Haugen to Sam Patten, March 28, 1924, Gilbert N. Haugen Papers, State Historical Society of Iowa, Iowa City.

[76] United States Congress, Senate, *Agricultural Export Bill*, Report 193, Part 2, 68th Cong., 1st Sess. Signing the minority report were Norris, Edwin F. Ladd of North Dakota, Magnus Johnson of Minnesota, and Arthur Capper of Kansas. Peter Norbeck wrote to a friend after the defeat of the McNary-Haugen bill: "I am frankly peeved over . . . the adroitness of Bob [Senator Robert M. La Follette of Wisconsin] . . . to divide the Northwest group on this important measure [McNary-Haugen bill] by advocating the impractical Norris-Sinclair bill, which everybody knew had no chance of decent support, even in the Northwest." Norbeck to A. C. Ellerman, July 22, 1924, Norbeck Letters, Box 1. James H. Sinclair of North Dakota introduced the companion bill in the House.

sentatives, particularly from the South, likewise undermined the chances of the bill. The crisis had not continued long enough nor had it affected all areas to a degree which would have united agricultural interests behind a single remedy.

Wallace was understandably disappointed with the defeat of the McNary-Haugen bill. In his more candid moments the secretary privately admitted to an uncomfortable uncertainty about the export corporation scheme. "I do not believe," he wrote to a former assistant, "any one can say with assurance that it will work one hundred per cent . . ."; in fact, he conceded, "how it will work out in practice I think no one can tell." But farmers were in a desperate state, and the McNary-Haugen bill appeared to be "the most hopeful of the plans which had been suggested. . . ." In Wallace's view, therefore, rejection of the measure had removed the last chance for the government effectively to aid agriculture through the depression. There remained only the movement of natural economic forces to bring about recovery, and that process was proving to be dangerously and painfully slow.[77]

The fight over the McNary-Haugen bill marked a critical turning point in Wallace's secretaryship. Endorsement of the export corporation plan represented not only the adoption of a new idea but a compromise of his basic philosophy regarding the relation of government to the economy. With the acceptance of Peek's scheme, Wallace moved perceptively to the left, nearer to Robert M. La Follette, Smith W. Brookhart, and the group which would shortly launch the Progressive party in the 1924 presidential campaign. This is not to say that the secretary became a radical or even that he arrived at a position consistent with that of the progressive bloc, but neither did he remain as conservative as he

[77] Wallace to C. W. Pugsley, January 8, 1924; Wallace to John D. Jones, Jr., January 25, 1924, Files of the Secretary of Agriculture; *Hearings on the Purchase and Sale of Farm Products*, p. 349; and *Wallaces' Farmer* 49 (January 18, 1924): 95.

had been in his first years in office. He also moved outside of that strand of progressivism which reached back to the administration of his idol, Theodore Roosevelt. However much conditions demanded it, a government-sponsored export corporation was something quite different from federal regulation of business or loans to farmers, a program of agricultural economics, or supervision of natural resources. Even as the Bull Moose candidate in 1912, Roosevelt would have had trouble fitting into his program a scheme in which the government was to enter the market to purchase and export goods, an equalization fee notwithstanding.

Wallace did not make this change unknowingly nor without careful consideration. In the midst of the fight over the McNary-Haugen bill he wrote to his brother that the existing farm situation demanded "a very determined effort to improve it, even if this effort involves doing some things which a few years ago we might not have been disposed to do." [78] Speaking of the laws he had helped to push through Congress in his first two years in office, he noted that they were "good but were rather a treatment of symptoms than of the disease," for they neither recognized nor attacked the price disparity which was at the root of the farm crisis. "It was not seen that agriculture in this difficulty needed help in an unusual way," Wallace explained shortly before his death, "and that unless such help were given, it would have to go through an unexampled period of liquidation, in which hundreds of thousands of farmers would be ruined." [79] Always sincere but never doctrinaire, the secretary decided that such a prospect was enough to justify acceptance of a plan which, although inconsistent with his basic philosophy, held out the possibility of avoiding catastrophe. Conditions forced him, in short, to pursue a goal

[78] Wallace to Dan A. Wallace, March 8, 1924, Files of the Secretary of Agriculture.
[79] Henry C. Wallace, *Our Debt and Duty to the Farmer* (New York: Century, 1925), pp. 102–3.

that would have sacrificed a degree of freedom for farmers in return for the survival and security of American agriculture.

Wallace's efforts in behalf of the McNary-Haugen bill virtually isolated him and his department from the rest of the administration. The relationship between the secretary and Coolidge was never cordial, but at least it had provided a basis for limited cooperation on routine matters. Disagreement over the export corporation plan, however, had destroyed even that slender link between the two men and rendered Wallace's continued presence in the cabinet impossible. During and after the fight over the McNary-Haugen bill, there were persistent rumors that Wallace was on the verge of resigning.[80] No doubt Coolidge would have welcomed such a move, for with the presidential election approaching he would have been reluctant to take the initiative himself and remove a man so popular with farmers. But the President was a near certainty to win the election, and it became clear that if Wallace had not voluntarily left the cabinet by the following March he would not win reappointment. The secretary's death in October 1924, then, saved Coolidge from the disagreeable task of passing over Wallace and assigning someone else to his post.

The months following the defeat of the McNary-Haugen bill confirmed Wallace's separation from the rest of the administration. In his acceptance speech for the Republican presidential nomination, Coolidge proposed a commission to investigate agricultural conditions and make recommendations to Congress. Representing the American Council of Agriculture, an organization formed to work for passage the McNary-Haugen bill after its initial defeat, Peek urged the President to place Wallace in charge of the commission. The Agriculture secretary himself on

[80] *Des Moines Register,* December 11, 1923, p. 1; *New York Times,* October 24, 1924, p. 12; and Taylor, "A Farm Economist in Washington," p. 313.

two different occasions made unsolicited suggestions on the makeup of the group. He wrote to Charles E. Hearst, head of the Iowa Farm Bureau, that he was "trying to bring about the appointment of a Commission that will be really helpful since one is to be appointed." [81] Agreeing with Hoover, upon whom he called for advice, that members of the commission must not be "infected by McNary-Haugen stuff," Coolidge rejected the requests of farm interests that Wallace be given a part in the planning. The final list contained not one of Wallace's choices, but a number of those recommended by the Commerce secretary.[82] Appointment of the commission shortly after Wallace's death was only the final piece of evidence demonstrating who, in fact, had been the real Secretary of Agriculture. As early as April the situation had prompted one disgruntled farmer to write to Coolidge: "We think his [Wallace's] advice rather than that of Mr. Hoover and his associates should be the administration's guide in matters affecting agriculture." Wallace saw more clearly than anyone the erosion of his influence and poignantly confided to a friend: "I only wish that I could make myself count more effectively on their [his supporters'] behalf." [83]

The presidential campaign gave further evidence of Wallace's loss of influence in the administration. Where four years earlier he had taken an active part in formulating the farm plank of the Republican platform and in directing Harding's campaign on

[81] George N. Peek to C. Bascom Slemp, September 3, 1924; Wallace to Coolidge, August 16, 1924; Wallace to Coolidge, September 21, 1924, Coolidge Papers, File 2180, Box 240; and Wallace to Charles E. Hearst, September 24, 1924, Charles E. Hearst Papers, University of Northern Iowa Library, Cedar Falls.

[82] Telegram, Rudolph Forster to C. B. Slemp, August 18, 1924; Hoover to Coolidge, August 29, 1924; list of those invited to join the commission, November 7, 1924, *ibid.*; telegram, Hoover to Coolidge, November 3, 1924; and telegram, Hoover to Coolidge, November 5, 1924, Hoover Papers, 1-I/241.

[83] M. C. Burritt to Coolidge, April 8, 1924, Coolidge Papers, File 1, Box 1; and Wallace to Charles E. Hearst, March 27, 1924, Hearst Papers.

matters concerning agricultural policy, in 1924 he did and said very little. Wallace made an innocuous statement to the effect that Coolidge was the "logical candidate of the Republican party" and that there was "no other candidate who could add more strength to the . . . party," but beyond this he was not asked nor was he inclined to go.[84] And although Wallace wrote three planks for the platform of the Iowa Republican party, one of which was an endorsement of the McNary-Haugen bill, the national platform revealed little of his influence.[85]

The McNary-Haugen issue, like the question over which department should have responsibility for the foreign marketing of farm products, became a factor in the selection of Wallace's successor. In August 1924 Chester C. Davis wrote to Taylor that "the defeat of the McNary-Haugen bill and the apparent vindication by Congress of President Coolidge's position . . . convinced me that the secretary of agriculture after March 4th next will not be a man with the same outlook on agriculture as that of Secretary Wallace." He anticipated that the post would go to someone "like Jardine of Kansas or Coulter of North Dakota." [86] As it happened, Davis turned out to be most prophetic. After Hoover's refusal to accept the office, Coolidge felt no hesitancy in offering the secretaryship to William M. Jardine. From his position as president of Kansas State College, Jardine had faithfully supported the administration's farm program and, most important, had spoken out against the McNary-Haugen bill. What could have been better from Coolidge's viewpoint than a Secretary of Agriculture from a leading wheat-producing state who opposed the export corporation scheme? If the President needed any more assurance that Jardine was the right man, Hoover provided it

[84] *New York Times,* October 11, 1923, p. 21.
[85] Telegram, Smith W. Brookhart to Coolidge, September 20, 1924, Coolidge Papers, File 2180, Box 240.
[86] Chester C. Davis to Taylor, August 7, 1924. Taylor Papers.

when he urged his appointment because he was "opposed to all paternalistic legislation. . . ." [87] When the new secretary took office in March 1925, he completed a solid administration front in opposition to McNary-Haugenism.

In the latter 1920's the McNary-Haugen scheme became the focus of attention of agricultural interests in their efforts to gain federal assistance. Changes in subsequent legislation to attract wider support and a decline in cotton prices led to the passage of reintroduced McNary-Haugen bills in 1927 and again in 1928. In each case, however, Coolidge vetoed the measure and farm representatives were unable to amass votes enough to override the President's action. Not until 1933, amidst the Great Depression and with a new administration in which Henry Agard Wallace was Secretary of Agriculture, did farmers see the enactment of legislation based upon the principle of parity for the prices of their products. Franklin D. Roosevelt's signing of the first Agricultural Adjustment Act in the first days of his Presidency marked the culmination of more than a decade of agitation by farm interests for a plan which Henry C. Wallace had done much to popularize.

[87] William M. Jardine to Hoover, April 14, 1924; telegram, Hoover to C. Bascom Slemp, November 3, 1924, Hoover Papers, 1-I/8 and 1-I 241. After his appointment, Jardine wrote to Hoover that he was "looking forward to a close association with you in my new responsibility." Jardine to Hoover, February 16, 1925, *ibid.*, 1-I/2.

Chapter XII / THE END
OF A CAREER

WALLACE'S last days in office were difficult ones. Failing health and the frustration of an increasingly isolated position within the cabinet caused him both physical and mental discomfort. Bothered with sciatica since early adulthood, he found the attacks becoming more frequent and more severe, until the condition would allow him to work no more than a few hours at a time and then only with considerable pain. A modest improvement in farm conditions by the middle of 1924, moreover, was small compensation for his disappointment over the defeat of the McNary-Haugen bill. And the continued support of many of the country's farmers represented an empty moral victory as he was being replaced by Hoover as their representative within the administration. Secretary Wallace had devoted three and a half trying years to the cause of American agriculture only to face removal from the cabinet once Coolidge took over the Presidency in his own right.

Certain that his days in public office were numbered, Wallace decided to write a book setting down his thoughts on the agricultural situation. Despite his long career in journalism, he felt more secure and was more effective delivering a speech than wielding a pen. But the kind of statement he wished to leave could not be made in a speech—it had to take a more permanent form. Thus, with the help of Nils Olsen of the BAE, Wallace undertook the task of writing his first and only book in the summer of 1924.[1]

Our Debt and Duty to the Farmer, published in 1925, advanced no new analysis of the farm problem nor did it offer anything different in the way of a solution. It merely brought together what Wallace had been preaching since the end of the war, under a title which pointedly expressed the essence of his feelings. In fact, large sections of the book were drawn directly from the annual reports, press releases, and other publications issued by the Agriculture Department during his secretaryship. The secretary dealt with the declining position of agriculture within the economy, the postwar crisis, and government efforts to aid the farmers. As he had many times in the preceding several years, Wallace held that disparity of prices was the principal factor contributing to the depressed condition of producers and insisted that the federal government was responsible for correcting the condition. The only question open to debate, as he saw it, was whether a fair price ratio might better be achieved by lowering the costs of goods and services which farmers needed or by raising the level of agricultural prices. Because of the economic instability and the financial hardship which would accompany a deflationary solution, Wallace considered the alternate approach the wisest. "By raising agriculture to the price plane upon which labor and industry are now seated so comfortably," he wrote, ". . . justice would be best served." This, of course, could be accomplished only through the disposition of agricultural surpluses, a purpose for which the proposed government export corporation was well adapted. Tediously moving toward its conclusion, *Our*

[1] Russell Lord, *The Wallaces of Iowa* (Boston: Houghton Mifflin, 1947), p. 256.

Debt and Duty to the Farmer arrived at the central point—a by-now-familiar defense of the McNary-Haugen bill.[2]

Work on his book nearly completed, Wallace's physician ordered him to the hospital for a minor operation to relieve the sciatic pain. Testing revealed, however, that his affliction was more serious than had originally been diagnosed, as it was discovered that part of the trouble was due to a gall bladder infection. An operation for the removal of both the gall bladder and appendix appeared at first to have been successful. But complications developed and he died 10 days later on October 25, 1924, at the age of 58. Secretary Wallace was given a state funeral, with the ceremony conducted, ironically, in the East Room of the White House, and Coolidge ordered an official day of mourning. Wallace's body was then transported back to Iowa where the family held its own services attended by hundreds of relatives, friends, and acquaintances.

Wallace's family claimed that the job in Washington had caused his untimely death. A martyr to the cause of agriculture was the way May Brodhead Wallace explained her husband's decease. More bitter and less inclined toward euphemism, Henry Agard declared unequivocally that Hoover had killed his father. To hold that the pressures and frustrations of the secretaryship had brought about an early death, however, is to take a rather sentimental and no doubt untenable view of the situation. Wallace was tough and hardy enough to endure his grueling three and a half years as head of the Agriculture Department. The operation was routine, and if it had not been for an unexpected case of intestinal poisoning there seems little doubt that he would have recovered.[3]

[2] Henry C. Wallace, *Our Debt and Duty to the Farmer* (New York: Century, 1925), pp. 189–208.

[3] Interview with Donald R. Murphy, Des Moines, Iowa, August 29, 1965.

Wallace's death was a shock to his subordinates in the Agriculture Department and to the country's farmers. "It was paralyzing news," Taylor later recalled. "I have never sat in a meeting where a group of intelligent men manifested such lack of power to think or act as in the meeting of Bureau Chiefs the next morning after the death of the Secretary." [4] Although not always in full agreement with his policies, farmers knew that they had a friend in Wallace, a friend who could be counted on to work for their interests. The death of the secretary alarmed them, therefore, the more so because of the generally unsympathetic attitude toward agriculture held by the rest of the administration. Many farmers would have agreed with Gifford Pinchot that "there was no one else in the Cabinet who could not have been spared very much more easily." [5]

As Secretary of Agriculture, Henry Cantwell Wallace was in some ways a success, in others a failure. Few men have occupied the chair who were as dedicated to the uplifting of American agriculture and to the goal of gaining fair treatment for the country's producers as he was. Reared in Iowa among farm people, he came to think as they did, embrace their values, adopt their prejudices, and apply their reasoning. He was ready to defend the interests of farmers with a determination and sincerity not seen in the previous administration and equalled earlier, perhaps, only by his close friend James Wilson. This dedication to those he was representing in Washington was one of his greatest strengths.

Not always sound in theory, Wallace was nevertheless instrumental in calling attention to an aspect of farming for which producers and their representatives had previously exhibited only limited concern. Since the latter nineteenth century, agriculture had been changing from a self-sufficient way of life to a commer-

---

[4] Quoted in Lord, *The Wallaces of Iowa,* p. 257.

[5] Pinchot to Edward L. Burke, November 6, 1924, Gifford Pinchot Papers, Manuscript Division, Library of Congress, Box 244.

cialized enterprise, thus becoming increasingly vulnerable to economic trends and fluctuations. But surprisingly few farmers were either aware of the changes taking place in their industry or of the new demands these changes were placing upon them. Although Secretary Wallace failed to understand the complexity of the movement, he possessed a greater appreciation of its problems than most of his rural friends. Despite the fact that it fell short of his objectives, the agricultural economics program went a long way toward developing wider awareness of the revolution that was modifying farming.

Perhaps one of Wallace's greatest successes as Agriculture secretary was in the administration of his department. A persevering worker and understanding chief, he earned the trust, admiration, and devotion of a loyal staff. Few governmental agencies at the time had the cohesion, the *esprit de corps* which existed in the Agriculture Department. Subjected to attacks from other departments, congressmen, the President, and outside critics, Wallace's division exhibited a stubborn tenacity and willingness to fight unusual even in bureaucratic Washington. To the secretary went the bulk of the credit, for through his own example he developed an atmosphere of mutual respect and purpose among his subordinates. On the other hand, his administrative success was tarnished by his failure to foster good relations with other members of the cabinet. Wallace was an outsider in an administration which was largely urban-oriented, well-educated, pro-business, and inclined to ignore the things he held most important. Never very diplomatic, he made little effort to understand his colleagues and in many instances increased his alienation from the rest of the cabinet by brash and arbitrary action. Wallace's unwillingness to compromise or cooperate with those in disagreement with him reduced his effectiveness as an administrator.

Because he believed that the farm depression was the result of dislocations caused by the war and thus temporary, Wallace seized upon solutions poorly suited to meet the situation. Higher tariffs, federal regulation, government assistance in the area of rural credit, and other piecemeal approaches were not addressed to the problems of an increasingly commercialized agricultural industry and failed measurably to stem the tide of the depression. Production adjustment was aimed more directly at the central weakness of the industry, but Wallace revealed a surprising naïvety about his agrarian friends when he insisted that such adjustment had to be on a voluntary basis. Nor was the export corporation plan any better designed to aid depressed farmers. The dumping of surpluses over tariff walls was a poor and inefficient substitute for production adjustment and an unrealistic vehicle for attacking the price disparity which had resulted primarily from a supply greatly in excess of domestic demand.

Wallace also lacked an appreciation of the plight of tenant farmers, who made up nearly 40 per cent of the agricultural population in 1920. A large proportion of this group, to be sure, suffered from essentially the same problems as landowning farmers and were in no worse condition. But many tenants, notably in the South, had special problems in addition to those of the agricultural community as a whole, which compounded their troubles and made their state even more precarious than that of farmers generally. Although the Agriculture Department under Wallace made a systematic study of tenancy, its efforts were largely directed toward improving the lot of independent landowners.[6] In neglecting a group so large as the tenant class, the secretary missed an opportunity to serve one of the most depressed elements in the country.

It would be unfair, however, to criticize Wallace too severely. He was Secretary of Agriculture during a difficult period, a period in which old problems were intensified and new ones appeared. Emerging from World War I, farmers became the victims

[6] USDA *Yearbook of Agriculture*, 1923, pp. 507–98.

of a convergence of difficulties which would have taxed the resources and energy of anyone who stepped into the post in 1921. If disastrous agricultural conditions were not enough, he also had to contend with a generally disinterested administration under Harding and an unsympathetic one under Coolidge. Had he been given a freer hand and greater support, he might have enjoyed a higher degree of success. While his farm program failed to solve the postwar crisis or to place agriculture in a better competitive position in relation to other sectors of the economy, federal assistance was important in helping producers through the critical period from 1920 to 1924 and in providing them with a measure of welcome relief, however inadequate for their needs.

Other proposals offered, moreover, were no better calculated than Wallace's to solve the problems of farmers. Inefficiency in marketing and distribution was a factor in high service costs and seasonal price depressions, certainly, and an improved system of cooperative marketing, as Hoover suggested, held out the possibility of reducing middleman charges and of regularizing the flow of goods to market. But as long as surpluses existed, benefits to producers from such a movement would have been small consolation in the face of continuing price disparity. Likewise, Coolidge's diversification proposal was not directed at agriculture's central problem. Outside of a few limited possibilities such as cotton and wool, there were no profitable alternatives to which farmers might have turned because of general overproduction throughout the entire agricultural industry. And, as Wallace pointed out, where adaptable at all diversification would have been a long-term process, and farmers needed immediate relief. Nor was price-fixing, though it would have temporarily relieved disparity, addressed to the basic weakness of agriculture. In fact, such a solution would have stimulated output and thus aggravated the very condition which was causing producers the most trouble. Norris' export corporation proposal would have reduced costs to farmers through government handling of foreign marketing, but, like the McNary-Haugen bill, would have left untouched the vital problem of adjustment in production and concentrated instead on dumping surpluses abroad. Designed to ease the effects of the farm situation rather than to correct it, these solutions were no more suitable than those advanced by Secretary Wallace.

Actually, the state of American agriculture in the 1920's demanded measures more far-reaching than anything seriously considered at the time. While acreage limitation schemes were cautiously suggested, no one ventured anything so radical as federal production control. Despite the obvious and broadly recognized fact that higher farm prices demanded a better balance between supply and demand, plans for compulsory restriction on output were not forthcoming. Since farmers had always been free to produce as much as they wished and had come to view their prerogative as something of a divine right, the absence of such proposals is understandable. They were willing to sacrifice a degree of their traditional independence in the name of security, but government adjustment of output was going too far. Such a solution would have to await a later day.

Yet, in the early 1920's farmers accepted, in fact they insisted on, a significant increase in government intervention in agricultural affairs as the only solution to their problems. As one of the leaders in the movement to secure federal help, Wallace had much to do with directing and shaping these demands. Like many agrarian spokesmen, the secretary initially advocated a limited program within well-defined areas, but with the failure of conditions to improve he was pushed toward proposals calling for greater and more direct federal interference in the economy. Through his role in the formation and work of the farm bloc, Wallace had a hand in the growing involvement of agriculture in politics. Under his secretaryship, furthermore, the Agriculture

Department fulfilled its traditional function as an advisory, service, and educational agency and at the same time became the leading lobby for agricultural legislation. In view of the farmers' growing dependence upon the federal government which has been developing since the early 1920's, Wallace takes on increased significance, for it was during his tenure as secretary that the movement was set firmly in motion.

Wallace influenced the formulation of subsequent farm policy in yet another and quite different way. Henry Agard Wallace followed his father's example not only in becoming editor of the family journal but later in accepting the post of Secretary of Agriculture in the Franklin D. Roosevelt administration. Just as Uncle Henry had influenced Henry C., so Henry C. passed on many of his traits to Henry A. A deep interest in securing justice for farmers, a militant temperament, a strong moral sense, an agrarian bias translated into resentment of urban values, an intensity of purpose, a narrow subjectivity in regard to certain matters—all of these characteristics were handed down and firmly implanted in the three generations of Wallaces. Henry C. Taylor once remarked that the Wallaces of Iowa were an institution, and a comparison of Uncle Henry, Henry C., and Henry A. would seem to bear out the contention.[7] Inasmuch as Henry C. shaped the ideas, values, and thinking of his son, he left a mark on the agricultural policy of the New Deal.

If Henry C. did not agree in all matters with his father, neither was he in complete accord with his son. As editor of *Wallaces' Farmer,* Henry A. endorsed the policy of the Agriculture Department under his father with one exception—the tariff—and on this issue the two men were at opposite poles. After 1924, however, Henry A. moved steadily to the left on the question of government intervention for the purpose of correcting the farm situ-

[7] Henry C. Taylor, "Henry C. Wallace," pp. 1–2, ms. in Henry C. Taylor Papers, State Historical Society of Wisconsin, Madison.

ation, and by the time he entered the Roosevelt cabinet in 1933 he was advocating federal restriction of production. Had Henry C. lived, the gulf between father and son would likely have widened, for it is doubtful that he could have accepted that much interference in the agricultural industry even during the Great Depression of the thirties. Another interesting speculation is whether he would have joined his son in leaving the Republican party to become a Democrat after Hoover's nomination as presidential candidate in 1928. Henry C. was a good Republican, certainly a more loyal one than Henry A., but for him to have accepted his old rival as President and leader of the party would be expecting a good deal. On the other hand, it would have been equally distasteful for him to have left his political home to join the Democrats. Perhaps he would have withdrawn from active participation in politics and supported neither party.

Wallace's career as Secretary of Agriculture was important on still another count. He was a central figure in a group of Iowans which exerted considerable influence at the national level around the turn of the century. Along with Jonathan P. Dolliver, Albert B. Cummins, and William S. Kenyon, Wallace sustained the tradition of reform politics symbolized by James B. Weaver and illustrated the kind of national leadership typified by John A. Kasson and William B. Allison. His position and activities, moreover, suggest several insights into what happened to the progressive movement between the end of World War I and the Great Depression and exemplify the thinking of a significant, if declining, faction within the Republican party.

On balance, Wallace met the postwar agricultural crisis as well as the country had a right to expect. Though his program fell short of the needs of farmers, it must be assessed in light of the times—times in which the power structure at the national level blocked attempts by him and others to carry out plans for more extensive federal assistance. That any really appropriate attack on agricultural problems would likewise have been rejected not only by the administration and Congress but by agrarians as well seems obvious. If Wallace may be criticized for faulty evaluation of the farm situation and for advancing inadequate solutions, he cannot be held responsible for failure to accomplish what in the 1920's would have been impossible.

Bibliographical Essay[*]

MANUSCRIPTS

Most valuable in the research for this work were the Files of the Secretary of Agriculture located in the National Archives. Not only do these files contain a vast amount of official communication relative to the work of the Department of Agriculture for the period 1921–1924, but they include as well Wallace's personal correspondence during his secretaryship. Also useful were hundreds of letters to the secretary from farmers and their spokesmen throughout the country in which they described agricultural conditions and presented the thinking of a representative cross-section of the agrarian community. The Files of the Bureau of Agricultural Economics, National Archives, were likewise an indispensable source, as they contain the pertinent information on the formation and activities of the department's most important and in many respects most controversial agency. For the fight between the Agriculture and Commerce Departments over foreign marketing work, the Files of the Secretary of Commerce and the Files of the Bureau of Foreign and Domestic Commerce, both in the National Archives, provided invaluable information. The Files of the Secretary of the Interior, National Archives, supplied much material on Albert B. Fall's efforts to gain control of the Forest Service and on the attitude and tactics of his department during the controversy over the transfer.

[*] See footnotes for a complete bibliography of the sources.

300

Wallace's personal papers, located at the University of Iowa Library, Iowa City, are fragmentary. On his activities as secretary of the Corn Belt Meat Producers' Association, his fight with Hoover during World War I, and his appointment to the Agriculture Department, the collection was particularly helpful. But, unfortunately, the files contain nothing on his years of practical farming nor on the formation of *Wallaces' Farmer,* and little from his secretaryship. As pointed out, however, the latter period is well covered in the Files of the Secretary of Agriculture. The files of the agriculture department and the reports of the Board of Trustees of Iowa State College, located at the University in Ames, provided material on Wallace as a professor.

By far the most valuable personal manuscripts investigated were the Henry C. Taylor Papers, State Historical Society of Wisconsin, Madison. Containing the bureau chief's correspondence, speeches by himself and others, office memoranda, newspaper clippings, and other items, the collection offers an excellent record of Taylor's and the Agriculture Department's activities in the early 1920's. Also in the papers are several manuscripts written by Taylor, the most important being "A Farm Economist in Washington, 1919–1925," which gives an account of his years in the department.

Two sets of presidential papers were utilized. The Warren G. Harding Papers at the State Historical Society of Ohio, Columbus, provided considerable information on the campaign of 1920, Wallace's appointment, and the secretary's efforts in behalf of agricultural legislation. Housed in the Manuscript Division of the Library of Congress, the Calvin Coolidge Papers were invaluable on the administration's efforts to undermine the McNary-Haugen movement and on Hoover's replacement of Wallace as the President's agricultural advisor in the cabinet.

The personal papers of two of Wallace's cabinet colleagues supplied pertinent material on important topics. Indispensable for the feud between the Agriculture and Commerce Departments and for the federal farm marketing board plan were the papers of Herbert C. Hoover, Herbert Hoover Presidential Library, West Branch, Iowa. The Albert B. Fall Papers, located at the University of New Mexico

Library, Albuquerque, contain several items on the Forest Service controversy. Barbara Anthes, *List of the Albert B. Fall Papers* (Albuquerque: University of New Mexico, 1957), describes each piece in the collection and enabled the writer to use the papers without going to Albuquerque.

Of some use were personal manuscripts of four congressmen. The incomplete Albert B. Cummins Papers, Iowa State Department of History and Archives, Des Moines, were helpful on the campaign of 1920 and to a lesser degree on the activities of Congress. Located in the Library of Congress, the papers of George W. Norris were most important on the history of his export corporation proposal in the early 1920's and on the farm bloc. The Gilbert N. Haugen Papers at the State Historical Society of Iowa, Iowa City, were disappointing, for they contain surprisingly little on the McNary-Haugen bill. The small collection of Peter Norbeck Letters, some originals but mostly typed copies, in the Western Historical Manuscripts Collection, University of Missouri Library, Columbia, was very useful for the activities of congressional agricultural representatives, particularly those in the farm bloc.

The George N. Peek Papers in the Western Historical Manuscripts Collection at Missouri were an important source in following the development of the Peek plan and the McNary-Haugen movement. Although the Chester C. Davis Papers, also in the Western Historical Manuscripts Collection, cover primarily the period 1925–1928, they supplied a few items on his activities in behalf of the McNary-Haugen bill in 1924. In regard to the conservationists' fight to retain the Forest Service in the Agriculture Department, the Gifford Pinchot Collection in the Library of Congress is the most important source. Of some use were the Edwin T. Meredith Papers at the University of Iowa Library and the Charles E. Hearst Papers at the University of Northern Iowa Library, Cedar Falls.

The files of the College of Agriculture of the University of Wisconsin, University of Wisconsin Archives, Madison, furnished material on Taylor's efforts to build an agricultural economics program at the school. Also instructive were Dean Harry L. Russell's correspondence relating to the connection between the extension service and the American Farm Bureau Federation.

GOVERNMENT PUBLICATIONS

Agriculture Department publications for the period of the early twenties offered considerable material. Particularly useful were the *Yearbooks of Agriculture,* which were prepared largely by the BAE and thus provide an index to the bureau's activities. In addition, *Annual Reports, Miscellaneous Publications, Bulletins,* and *Agricultural Situation* proved useful. Publications of the Department of Commerce, particularly the "Survey of World Trade in Agricultural Products" series in *Trade Information Bulletins,* and the *Annual Report of the Secretary of Interior* for 1921 and 1922, shed some light on the interdepartmental controversies during Wallace's tenure. In connection with the secretary's interest in bringing about reduction in railroad rates, the *Interstate Commerce Commission Annual Report,* 1921–1924 supplied supplemental information. For following the history of agricultural legislation, the records of Congress were of course most helpful. The *Congressional Record,* records of various hearings, committee reports, and congressional documents revealed much on the efforts of farm representatives and the activities of the Agriculture Department in regard to agricultural bills.

INTERVIEWS

Donald R. Murphy, who lives in Des Moines, has been associated with *Wallaces' Farmer* since 1918 and was a personal friend of both Wallace and his oldest son. An interview with Murphy on August 29, 1965, was most instructive, as he was able to furnish a good deal of material not available in manuscript or printed sources. Of great interest and value were Murphy's perceptive insights into Wallace's personality and his relating of anecdotes about the secretary. In an interview on December 21, 1966, Wallace's son James, also of Des Moines, provided background on the family and personal comments on his father. In addition, he made available Wallace family memorabilia in his possession.

MEMOIRS AND AUTOBIOGRAPHIES

For background material, Henry Wallace, *Uncle Henry's Own Story of His Life,* 3 volumes (Des Moines: Wallace, 1917–19) was an excellent source. David F. Houston, *Eight Years with Wilson's Cabinet,* 2 volumes (New York: Doubleday, Page, 1926) gives an account of the activities of the Agriculture Department under Wilson's administration. In connection with the fight between Wallace and Hoover, *The Memoirs of Herbert Hoover,* 3 volumes (New York: Macmillan, 1952) supplied several interesting and enlightening items. On agricultural legislation and the work of the farm bloc, see Arthur Capper, *The Agricultural Bloc* (New York: Harcourt, Brace, 1922) and George W. Norris, *Fighting Liberal* (New York: Macmillan, 1945). Frustratingly incomplete but occasionally useful was *The Autobiography of Calvin Coolidge* (New York: Cosmopolitan Book Corporation, 1929). Useful for other special topics were Harry M. Daugherty, *The Inside Story of the Harding Tragedy* (New York: Churchill, 1932), Samuel Gompers, *Seventy Years of Life and Labor,* 2 volumes (New York: E. P. Dutton, 1925), William B. Greeley, *Forests and Men* (Garden City: Doubleday, 1951), and Hugh S. Johnson, *The Blue Eagle from Egg to Earth* (Garden City: Doubleday, Doran, 1935).

NEWSPAPERS AND PERIODICALS

Wallaces' Farmer was, of course, a basic source. For the years under the editorship of Henry C., the journal provided insights into his thinking and a good account of his activities. Since Wallace took no part in formulating editorial policy after he became Secretary of Agriculture, the paper was less useful for the early twenties. Nevertheless, the stand taken by *Wallaces' Farmer* under the guidance of Henry Agard was interesting and often enlightening.

Besides *Wallaces' Farmer,* several other farm journals were useful, particularly *Prairie Farmer* and *Progressive Farmer.* Publications of agricultural organizations, such as the *American Farm Bureau Federation News Letter, Iowa Union Farmer,* and *The Producer,* likewise afforded some items of importance. For the attitude of Gompers on the National Agricultural Conference see *American Federationist,* and for the conservationist campaign against the Forest Service transfer see *American Forestry.*

Of the newspapers investigated, three were particularly valuable. Wallace's hometown sheet, *Des Moines Register,* followed his career closely and frequently gave him more complete coverage than the larger papers. Close to the scene of government activities, the *Washington Post* was often able to report on the activities of the Agriculture Department and congressional farm representatives in greater depth and detail than outside newspapers. For accurate, dependable, and unbiased accounting, the *New York Times* as always was the best. In addition, two other Iowa newspapers, the *Fort Dodge Messenger* and the *Des Moines Capital,* were helpful, as well as the *Chicago Tribune, Christian Science Monitor, New York World,* and *Washington Herald.*

SECONDARY SOURCES

The best and most comprehensive study of agricultural history in the early 1920's is James H. Shideler, *Farm Crisis, 1919–1923* (Berkeley and Los Angeles: University of California Press, 1957). Broader in scope and thus less intensive are Gilbert C. Fite, *George N. Peek and the Fight for Farm Parity* (Norman: University of Oklahoma Press, 1954), Theodore Saloutos and John D. Hicks, *Agricultural Discontent in the Middle West, 1900–1939* (Madison: University of Wisconsin Press, 1951), Murray R. Benedict, *Farm Policies of the United States, 1790–1950* (New York: Twentieth Century Fund, 1953), and Fred A. Shannon, *American Farmers Movements* (Princeton: D. Van Nostrand, 1957). The most recent history of the Agriculture Department is Gladys Baker, Wayne D. Rasmussen,

Vivian Wiser, and Jane M. Porter, *Century of Service: The First 100 Years of the United States Department of Agriculture* (Washington: USDA, 1963). For the development of farm economics and the formation and activities of the BAE, Henry C. and Anne D. Taylor, *The Story of Agricultural Economics, 1840–1932* (Ames: Iowa State College Press, 1952) is the definitive work. Wallace's book, *Our Debt and Duty to the Farmer* (New York: Century, 1925), is a good synthesis of his views and farm program.

A number of biographies supplied valuable supplemental material: Russell Lord, *The Wallaces of Iowa* (Boston: Houghton Mifflin, 1947); Andrew Sinclair, *The Available Man: The Life Behind the Masks of Warren Gamaliel Harding* (New York: Macmillan, 1965); Claude M. Fuess, *Calvin Coolidge: The Man from Vermont* (Boston: Little, Brown, 1940); Gilbert C. Fite, *Peter Norbeck: Prairie Statesman* (Columbia: University of Missouri Press, 1948); Claudius O. Johnson, *Borah of Idaho* (New York: Longmans, Green, 1936); Belle C. and Lola La Follette, *Robert M. La Follette,* 2 volumes (New York: Macmillan, 1953); Alfred Lief, *Democracy's Norris* (New York: Stackpole Sons, 1939); Homer E. Socolofsky, *Arthur Capper: Publisher, Politician, Philanthropist* (Lawrence: University of Kansas Press, 1962); Merlo M. Pusey, *Charles Evans Hughes,* 2 volumes (New York: Macmillan, 1951); M. Nelson McGeary, *Gifford Pinchot* (Princeton: Princeton University Press, 1960); William T. Hutchinson, *Lowden of Illinois,* 2 volumes (Chicago: University of Chicago Press, 1957).

Index

Agricultural Adjustment Act, 288
Agricultural conditions: later nineteenth century, 16, 21, 61-62; early twentieth century, 62-63; World War I, 63-64; postwar period, 64-65; early 1920's, 65-70, 247
Agricultural Credits Act, 83-86
Agricultural Credit Corporation, 273-274, 274*n*
Agricultural extension: and American Farm Bureau Federation, 136-140; and BAE, 136-140; in Pennsylvania, 137*n*; in Wisconsin, 137*n*
Agricultural Marketing Act, 125
Allison, William B., 39, 298
American Bankers' Association, 268
American Cotton Association, 120-122. *See also* Wannamaker, J. S.
American Farm Bureau Federation: support of Wallace for Secretary of Agriculture, 53; and farm bloc, 72-74; and Fordney-McCumber Tariff, 97; and agricultural extension, 136-140 *passim;* and agricultural conference, 148; and foreign marketing of agricultural products, 229*n;* and Ketcham bill, 240*n;* and Wins-

low bill, 240*n;* and Peek plan, 255-256; and McNary-Haugen bill, 269; and Capper-Williams bill, 278; mentioned, 47, 203
American Forestry Association, 169, 172-173
Anderson, Sydney: and agricultural credit, 101; as chairman of agricultural conference, 151; and Peek plan, 155; and Bureau of Markets, 229
Armour, J. Ogden, 203-208 *passim*
Aswell, James B., 281

Barnes, Julius H.: opposition to Peek plan, 254-255; opposition to McNary-Haugen bill, 268; mentioned, 264
Baruch, Bernard M.: and agricultural conference, 157-158; and Peek plan, 158, 268; support for McNary-Haugen bill, 268; mentioned, 248
Bradfute, O. E., 277-278
Brand, Charles J.: and Armour-Morris merger, 210-212; and foreign marketing of agricultural products, 221; and Peek plan, 267; and McNary-Haugen bill, 279
Brookhart, Smith W.: opposition to

307

A Note on the Author

Donald L. Winters is associate professor of history at the University of Northern Iowa, Cedar Falls. He received his B.A. from that university in 1957 and his Ph.D. from the University of Wisconsin in 1966. *Henry Cantwell Wallace as Secretary of Agriculture, 1921–1924,* his first book, is the 1969 winner of the Agricultural History Society award.

UNIVERSITY OF ILLINOIS PRESS